D1270505

Wireless Communications
Evolution to 3G and Beyond

For a listing of recent titles in the *Artech House Mobile Communications Library,* turn to the back of this book.

Wireless Communications Evolution to 3G and Beyond

Saad Z. Asif

ARTECH
HOUSE

BOSTON | LONDON
artechhouse.com

Library of Congress Cataloging-in-Publication Data

British Library Cataloguing in Publication Data

ISBN-13: 978-1-59693-022-3

Cover design by Yekaterina Ratner

© 2007 ARTECH HOUSE, INC.
685 Canton Street
Norwood, MA 02062

We acknowledge the following sources for providing copyright permission to reprint material in tables and figures in this book:

CCSA;
CDMA Development Group;
Com GC Press, Siemens;
ECMA International;
European Telecommunications Standards Institute (ETSI);
Freescale Semiconductor, Inc.;
GSM Association;
Hot Telecom;
IEEE;
International Journal of Networking Security (INJS);
John Wiley & Sons, Inc./ John Wiley & Sons, Ltd.;
The McGraw-Hill Companies;
Microwave Journal;
Pearson Education;
Qualcomm Incorporated;
Signals Research;
Springer/Kluwer;
Tektronix, Inc.;
Telecommunications Industry Association (TIA);
Telecommunications Technology Association (TTA);
World Wireless Congress.

10 9 8 7 6 5 4 3 2 1

*To my beloved parents, Muhammad and Nusrat,
to my siblings, Babar and Sarah,
to my dear wife, Saira, and to our lovely daughters,
Maha and Shiza*

Contents

Preface

In the name of Allah, the Most Merciful, the Most Compassionate

I usually plan things well ahead of time both in personal and professional lives. Contrary to that normal approach, my 5-year plan hadn't included writing a book. Quite unexpectedly, I contacted Artech House and things start moving in the right direction in late 2004. By the grace of God, I completed my first book in 2006. I hope you will enjoy reading this book and reap immense benefits from it.

The book is intended for everyone who is involved and interested in wireless communications including engineering managers, engineers, researchers, students, and telecommunication professionals. More specifically, I would hope that developing nations who lack development in the areas of third generation (3G) and wireless R&D will benefit from this book. Thus, my aim as a resident of a developing nation (Pakistan) who has also spent many years conducting telecommunication R&D activities in the United States is to encourage the audience in the developing nations to look beyond acquiring the technologies and systems from the developed nations. Instead, these countries should start conducting R&D work, participating in the standard bodies and submitting proposals, collaborating with other developing nations, building manufacturing and R&D houses, and building a sound technological foundation for the future generations of their respective countries.

This book takes an in-depth look at current 3G and broadband wireless access technologies, provides some insights for the future of these networks, and paves the path toward 4G. In addition, it also looks into local-area and personal networks. Interoperability aspects between the various technologies, deployment aspects for 3G technologies, and security features of these systems are also discussed. In the end, it provides the different sets of applications that could be useful for businesses and consumers and describes the market conditions for wireless technologies for some developed, emerging and developing countries.

Chapter 1 provides a summary of the overall book. Chapter 2 illustrates the key concepts and the process of information flow, briefly describes the key features of 1G and 2G technologies, and presents the evolution (both network and devices) toward 3G.

In Chapters 3, 4, and 5, the 3G technologies of CDMA2000, UMTS, and TD-SCDMA are described. Chapter 3 provides details of CDMA2000 1X technology and its migration path including 1xEV-DO and 1xEV-DV technologies. It looks into the network topologies of 3G1x and EV-DO and briefly outlines the future of CDMA2000. Chapter 4 provides details of the WCDMA access technology and its evolution to HSDPA and features of Release 4/5. Chapter 5 provides an overview of the TD-SCDMA standard, which is homegrown Chinese technology. It also provides a detailed description of the three layers of the TD-SCDMA radio interface and the integration processes of TD-SCDMA with GSM/GPRS and UMTS networks.

Chapter 6 looks into wireless local loop (WLL), its architecture, and TDMA-based WLL technologies such as DECT, PACS, and PHS. It also describes WCDMA- and CDMA2000-based WLL systems and also provides market statistics for the WLL systems. Chapter 7 discusses the broadband technologies, including WiMAX, that are targeted to solve the last mile problem.

Chapter 8 provides an overview of WLAN and WPAN technologies including IEEE 802.11 series, Bluetooth, and HIPERLAN2. In Chapter 9, mobility management for 3GPP- and 3GPP2-based systems are discussed. It also looks into the interoperability aspects between 3G technologies and between cellular and WLAN technologies.

Chapter 10 highlights the process from inception to deployment for the wide-area wireless technologies (e.g., 3G). Chapter 11 briefly describes 3G, WiMAX, WLAN, and WPAN radio access security measures. It also addresses the threats to access network and possible mitigations and provides an analysis on end-to-end (wireline and wireless) security.

Chapter 12 briefly defines the developments that are in progress for systems beyond 3G including features that would make these technologies more appealing. It also presents a high-level action plan that could be implemented to avoid further bifurcations among various standard bodies for 4G. Chapter 13 discusses a number of applications for the business and consumer segments of the market. It also presents the telecommunication developments for developing, emerging, and developed nations. A lot of growth is expected in the wireless telecom markets of Arab world, South Asia, China, and India, whereas market saturation is expected in Europe and the United States in the coming years.

From the bottom of my heart I would like to thank my wife for her everlasting support, my parents for their undying support and never-ending prayers, and my brother for his assistance in proofreading.

CHAPTER 1

Summary and Introduction

This book provides details about the various types of wireless technologies that are available today. Furthermore, it reviews the short-term and long-term prospects for these state-of-the-art wide-area, metropolitan-area, local-area, and personal-area wireless technologies.

1.1 What We Have Today

Today we have multiple competing technologies and standards in the cellular world. The Global System for Mobile Communications (GSM; see Table 1.1 for a list of the acronyms used throughout) has, by far, the largest subscriber base with more than 1.5 billion users, whereas CDMA has close to 320 million subscribers worldwide. CDMA is mostly used in North America and in Asian, Latin American, and Caribbean countries. CDMA has not had much success in Europe, Middle East, and Africa compared to GSM usage in these parts of the world. The majority of the emerging and developing markets are still using 2G (GSM/GPRS and CDMAOne) technologies, whereas a number of countries in the developed world have already deployed 3G solutions.

The GSM operators have paid billions of dollars to acquire 3G/WCDMA licenses but they haven't fully recovered the price tag yet. On the other hand, CDMAOne operators did not need a separate 3G license, and migration to CDMA2000 was relatively easy compared to the transition from GSM to WCDMA. The WCDMA and CDMA2000 migration paths were standardized by 3GPP and 3GPP2 committees, respectively.

China, which has the largest number of cellular subscribers, has developed its own 3G standard called TD-SCDMA but has not so far issued 3G licenses. TD-SCDMA is a distinct but unproven 3G technology that uses the GSM core network to minimize the cost of upgrade for the operators of the world's largest GSM base. TD-SCDMA, a homegrown technology, was standardized through the efforts of CCSA within Release 4 of 3GPP. Thus, one key common item among all three key 3G standards (CDMA2000, WCDMA, and TD-SCDMA) is that they are all based on the CDMA technology.

In recent years, the world has also seen the entry of the IEEE into the cellular (mobile) world. The most talked about entry is the WiMAX technology developed by the IEEE 802.16 working group to provide broadband wireless access. The initial focus of this group was on fixed broadband wireless but now with the standardization of the 802.16e standard, the IEEE has started to compete directly with mobile broadband technologies that include EV-DO (Release 0/A/B) of 3GPP2 and HSPA (HSDPA and HSUPA) of 3GPP. In addition, the IEEE 802.20 (Mobile Broadband Wireless Access) committee was brought back to life with an OFDM-based proposal on mobile broadband wireless access to compete with WiMAX.

Beyond cellular and WiMAX technologies, choices include WLL and WLANs. The CDMA2000 WLL deployments are constantly on the increase, especially in developing nations where the cost of basic landline telephony service is very cost prohibitive. The WLAN market is also growing slowly and interoperability aspects with GSM and CDMA are widely entertained. The IEEE 802.11b/g standard is the most popular WiFi solution available in the marketplace. In addition, we have WPANs such as Bluetooth to enable short-range wireless connectivity with personal devices.

1.2   What We Can Expect in the Short Term

The two near-term key expectations are issuance of 3G licenses in China and a few announcements in favor of WiMAX from 2G (maybe one or two from 3G) operators. The operators in China are expected to initially receive only one 3G license (i.e., TD-SCDMA) with other 3G licenses awarded at a later date. Some GSM operators are focusing more on WiMAX than on 3G because of the heavy price tag that is required to acquire 3G licenses.

In the near future, four 3.5G technologies are also expected to be available for the mobile workforce. The 3G-LTE, EV-DO (Release C), IEEE 802.16e (when it will support full mobility), and IEEE 802.20 technologies are considered to be 3.5G technologies in this book. The evolution of CDMA2000 to EV-DO (Release C), of HSPA to 3G-LTE, and the rebirth of IEEE 802.20 are in direct competition with IEEE 802.16e. All four of these technologies are technically very similar and all use OFDMA, unlike 3G technologies, which are all based on CDMA.

WiMAX (IEEE 802.16e) could appear on the road map for many 2G operators because the spectrum for WiMAX is relatively much cheaper than the 3G licenses. Although there is no direct interaction between GSM and WiMAX, it would still be cheaper to deploy WiMAX instead of migrating from GPRS/EDGE to UMTS. On the other hand, the 3G operators are still debating the validity of WiMAX and tend to lean more toward 3G-LTE and EV-DO (Release C).

The WLAN IEEE 802.11n standard is expected to provide much higher data rates and is likely to be completed by mid-2007. The failure of IEEE 802.15.3, a task group formed to decide between the DS-UWB technology of the UWB Alliance and the MB-OFDM technology of the WiMedia Alliance, may eventually lead to two different sets of specifications for the next generation of WPANs.

1.3   What We Can Hope for in the Long Term

The evolution to 3G was envisioned by ITU under the IMT-2000 project. The IMT-2000 project was started with 3GPP and 3GPP2 committees for the standardization of 3G. 3GPP looked after the evolution of the GSM systems, whereas 3GPP2 was launched to complement the evolution study of non-GSM systems such as CDMAOne. Since their inception the two groups have made steady progress and it was hoped that at some point in time they would converge.

Instead of converging we now have three additional groups to study the evolution of mobile wireless broadband, which makes the picture even more complex. These groups are IEEE 802.16, IEEE 802.20, and CCSA, which are standardizing WiMAX, MBWA, and TD-SCDMA, respectively.

What is presented in this book is an action plan for ITU to consider in order to avoid further divisions within the world of standardization for 4G. 4G is still undefined but the convergence of cellular and IP is happening and will be a key piece of the puzzle for future networks (Figure 1.1). Offering newer applications will be the key differentiation point among operators rather than the underlying technology. Thus, defining multiple standards will not be so effective and will not generate economies of scale. The customers are more interested in applications that can help them in effectively managing their tasks related to mobile devices rather than the underlying technology.

The 4G standard should provide a smooth migration for the 3G/3.5G systems. The telecommunication industry has invested billions of dollars in 2G/3G networks and without a good return on investment (ROI), the operators and manufacturers will not make a serious move toward 4G. The convergence will expedite the commercial availability of the products, reduce costs, and offer seamless mobility. The main task of 4G or convergence should be to bind the operators' wide-area network with the local-area networks, thus enabling seamlessness for customers inside and outside of the home for all applications and services without tying up their knots with WiFi.

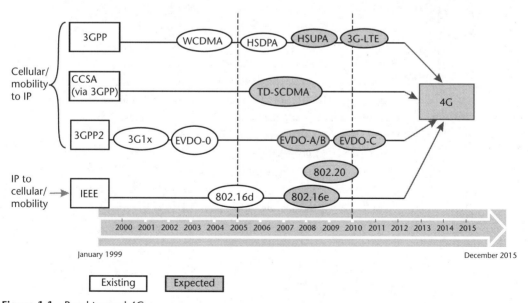

Figure 1.1   Road toward 4G.

1.4 Book Overview

In a nutshell, this book looks into various wireless access technologies for personal-, local-, metropolitan-, and wide-area networks. It also describes the deployment aspects of 3G cellular technologies and security aspects of multiple wireless technologies. It then describes the evolution of technology beyond 3G, in which seamless mobility and broadband may be considered as key requirements for harmonization among various standards bodies for 4G. In the last chapter it describes the applications for the business and consumer segments of the market and also provides an update about the overhaul of the cellular industry that is currently taking place in many developing countries around the world. Thus, the book can be divided into the following broad areas (see Table 1.1):

Table 1.1   Acronyms

Acronym	Description
2G	Second generation
3G	Third generation
3G-LTE	3G–long-term evolution
3GPP	Third Generation Partnership Project
3GPP2	Third Generation Partnership Project 2
4G	Fourth generation
CDMA	Code division multiple access
CCSA	China Communications Standards Association
DS-UWB	Direct spread-UWB
EV-DO	Evolution data optimized
GPRS	General Packet Data Radio System
GSM	Global System for Mobile Communications
HSDPA	High-speed downlink packet access
HSPA	High-speed packet access
HSUPA	High-speed uplink packet access
IEEE	Institute of Electrical and Electronics Engineers
IMT-2000	International Mobile Telecommunications 2000
ITU	International Telecommunications Union
MB-OFDM	Multiband OFDM
MBWA	Mobile broadband wireless access
OFDMA	Orthogonal frequency division multiplexing access
TD-SCDMA	Time division-synchronous CDMA
UMTS	Universal Mobile Telecommunication System
UWB	Ultrawideband
WCDMA	Wideband CDMA
WiFi	Wireless fidelity
WLAN	Wireless local-area network
WLL	Wireless local loop
WPAN	Wireless personal-area network
WiMAX	Worldwide interoperability for microwave access

- Chapters 2 to 8 look into various wireless access technologies (*what we have today and what is expected in the short term*).
- Chapter 9 discusses the interoperability aspects among various wireless technologies.
- Chapter 10 explores the deployment aspects of 3G technologies.
- Chapter 11 takes a deeper look at the security aspects of wireless access technologies.
- Chapter 12 looks beyond 3G (*what we can hope for in the long term*).
- Chapter 13 briefs the reader on the applications for businesses and consumers. It also highlights the prevailing market conditions and those that are expected in future in different parts of the world.

CHAPTER 2
Mobile Wireless Communications Evolution: 1G, 2G, and Beyond

2.1 Introduction

This chapter briefly looks into the three generations of mobile wireless communications: first generation (1G), second generation (2G), and third generation (3G). 1G systems were analog and evolved during the early 1980s, 2G systems were made available in the late 1980s with the advent of digital technology, and 3G systems evolved in the late 1990s to provide packet data services. This chapter also illustrates some key attributes of the mobile systems, the key blocks of the digital communications that are applicable to 2G and 3G technologies, high-level information on some key 2G systems, and the evolution of the 3G systems.

2.1.1 1G Cellular Systems

The first generation of cellular wireless communications was based on analog technology and was made available during late 1970s and early 1980s. The 1G analog systems only supported circuit-switched (dedicated line) voice calls. Some of the most successful 1G systems are listed here:

- *Nordic Mobile Telephone:* Later in Europe, the Nordic Mobile Telephone system (NMT 450) was developed in 1981 for the 450-MHz band and uses 25-kHz channels.
- *AMPS:* In 1983, the U.S. Federal Communications Commission allocated 666 duplex channels (40 MHz of spectrum in the 800-MHz band, with each channel having a one-way bandwidth of 30 kHz for each duplex channel) for the U.S. Advanced Mobile Phone System (AMPS).

2.1.2 2G Systems

The growth of digital communications throughout the world in the late 1980s allowed the cellular industry to develop its own second generation digital cellular systems (DCSs). The four key DCSs are as follows:

- *Global System for Mobile Communications (GSM):* GSM is the most popular 2G digital cellular standard in the world, first deployed in 1991 in the 800/900-MHz band. Details about GSM are provided in the next section.

- *IS-54:* In late 1991, the first U.S. digital cellular system or Interim Standard IS-54 hardware was installed in major U.S. cities. This system allows cellular operators to replace some single-user analog channels with digital channels that support three users in the same 30-kHz bandwidth. Details are provided in the next section.
- *IS-95:* In 1993 a cellular system based on code division multiple access (CDMA) was standardized by Telecommunications Industry Association (TIA) as Interim Standard 95 (IS-95). Details are provided in the next section.
- *PDC:* Pacific Digital Cellular (PDC) technology was developed in Japan. It is based on TDMA and details are provided in the next section.

The advent of 2G also included the personal communications system (PCS) frequency band, which operates around 1900 MHz, in addition to the 800-MHz band. There are many similarities between the systems of the 800- and 1900-MHz frequency bands, as well as differences. Besides the frequency band difference, the other major differences are in terms of coverage and in-building penetration between the two bands. The 2G digital systems (both cellular and PCS) not only support circuit-switched voice calls but also circuit-switched data communications.

2.1.3 3G Systems

The ITU (International Telecommunications Union) in the late 1990s saw the growth of the Internet and other packet data applications as the key to developing third generation cellular systems. These systems not only support legacy applications but also packet data services. The details of 3G systems are provided in Chapters 3, 4, and 5.

2.2 Key Attributes of Cellular/PCS Technologies

This section covers some of the basic concepts and architectures of 2G standards. These concepts and network elements in most cases are part of both 2G and 3G technologies.

2.2.1 2G Architecture

Figure 2.1 depicts the network elements of a typical 2G wireless network. Let's first define the common terms and network elements that are essential to understanding the systems' concepts [1, 2].

- *Air interface:* The air interface is the radio-frequency portion of the circuit between the mobile phone and the base transceiver station.
- *Authentication center (AUC):* The AUC verifies the identity of a user before granting permission to provide service to that user. It does so by processing the authentication response sent by the user.
- *Base station:* A base station has two components: the base transceiver station (BST) and the base station controller (BSC). The BTS consists of radio

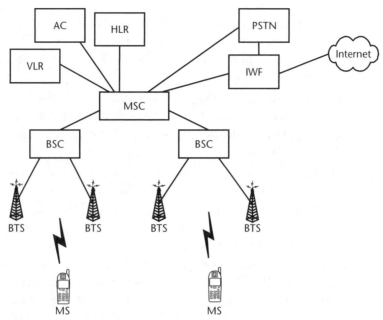

Figure 2.1 2G wireless network.

channels and transmitter and receiver antennas that are mounted on the tower. A BSC can serve multiple BTSs and provides call processing.

- *CDMA:* In CDMA, a frequency channel is used by multiple subscribers in a given coverage area and the signals are distinguished by spreading them with different codes.
- *Cell:* A cell is a specific geographic area covered by a BTS.
- *FDD:* In frequency division duplex (FDD) a frequency channel has two different frequencies, one for the transmitter and one for the receiver.
- *FDMA:* In frequency division multiple access (FDMA) users are assigned a channel from a limited number of channels ordered in the frequency domain. In FDMA systems once a frequency is assigned to a user, the designated frequency is exclusively used by the subscriber until he or she no longer needs the resource.
- *Forward link or downlink:* The set of communications from the base stations to the mobile stations.
- *Home location register (HLR):* This is a database that contains subscriber information.
- *Interworking function (IWF):* The IWF enables the circuit-switched data service in 2G networks.
- *Mobile station (MS):* The MS is a wireless device used by the subscriber to access other mobile/landline users and data services.
- *Mobile switching center (MSC):* The MSC switches user traffic that travels between the MS and the public switched telephone network or between the MS and another MSC. A MSC can serve multiple BSCs.

- *Network management system (NMS):* The NMS simply performs operations, administration, maintenance, and provisioning (OAM&P) functionalities.
- *Reverse link or uplink:* The set of communications from the mobile stations to the base stations.
- *Roaming:* A mechanism that allows mobile stations to operate in a service area other than that from which service has been subscribed.
- *Sector:* A cell can be divided into multiple sectors, each using the same BTS but their own separate antennas.
- *TDD:* Radios in time division duplex (TDD) systems transmit and receive on the same frequency using different timeslots.
- *TDMA:* In time division multiple access (TDMA), the usage of each ratio channel is partitioned into multiple timeslots, where all users use the same frequency but different timeslots.
- *Visitor location register (VLR):* This is a database that contains selected information from the HLR. This information is necessary for call control and provision of the subscribed services for each mobile currently located in the geographical area controlled by the VLR. It also provides links to one or more MSCs.

2.2.2 Key Concepts

Mobile communications, whether cellular or Personal Communication Systems (PCS), is a form of wireless communications in which some of the key concepts discussed next are utilized.

2.2.2.1 Handoffs

Cellular systems use two main types of handoffs: soft and hard. Soft handoffs transfer calls before disconnecting them (i.e., make before break), whereas hard handoffs transfer calls with interruption (i.e., break before make). Soft handoffs usually take place between base stations of the same MSC, softer handoffs between sectors of the same BTS, and hard handoffs between different MSCs.

2.2.2.2 Trunking and Grade of Service

Grade of service (GOS) is normally defined as the probability that a call is blocked, or the probability of a call experiencing a delay greater than a certain queuing time.

The measure of traffic intensity bears the name of a Danish mathematician Agner Erlang who developed the fundamentals of the trunking theory.

2.3 Digital Communications Systems

The development of low-rate digital speech coding and continuous improvements in the application specific integrated circuits (ASICs) paved the way for 2G, 2.5G, and 3G digital cellular systems. The digitization also allowed the use a variety of multiple-access techniques, namely, CDMA and TDMA as alternatives to FDMA.

Digital systems can support more users per base station and per allotted frequency spectrum, allowing wireless service providers to operate more economically.

2.3.1 Digital Communications System Blocks

Because 2G and 3G systems are all based on digital technology it is important to understand the basic flow of information in digital communications systems (Figure 2.3). The 2G and 3G systems use major processes of source coding, channel coding, and modulation to enable voice and data services in a more effective way than the 1G analog systems.

Figure 2.2 illustrates the basic elements of a digital communication system (DCS). The source output may be an either an analog signal, such as audio or video signal, or a digital signal. The key steps of speech coding, channel coding, and modulation are discussed next.

2.3.1.1 Vocoding

In DCS, the messages produced by the source are usually converted into a sequence of binary digits, which is the first step of the process. The goal is to seek an efficient representation of the source output that results in little or no redundancy. The process of converting the output of either an analog or a digital source into a sequence of binary digits is called *source encoding*. In mobile communications systems, vocoders or source coders achieve very high transmission bit rate reductions and are based on using appropriate knowledge about the signal to be coded; they are also signal specific. Also, the type of multiple-access technique (FDMA, TDMA, or CDMA) used is an important factor in determining the spectral efficiency (bit per second per hertz) of the system, and strongly influences the choice of speech coders.

In the past, GSM used the regular pulse excited–long-term prediction (RPE-LTP) scheme. In this scheme, the information from previous samples, which

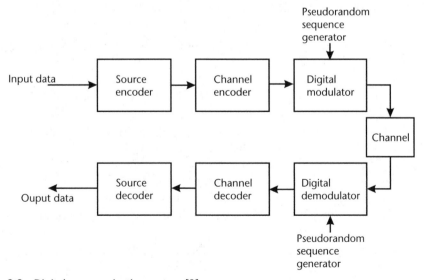

Figure 2.2 Digital communication system [3].

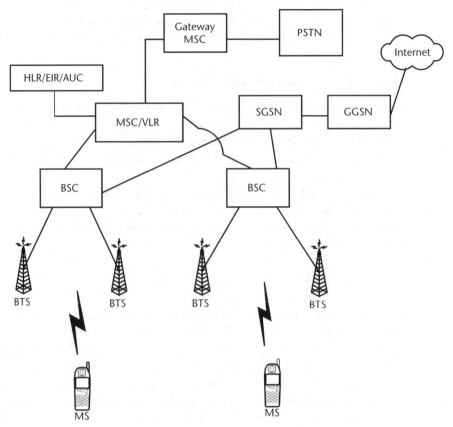

Figure 2.3   GPRS network architecture.

does not change very quickly, is used to predict the current sample. The coefficients of the linear combination of the previous samples, plus an encoded form of the residual, the difference between the predicted and actual sample, represent the signal. IS-54/IS-136 uses the 8-Kbps vector sum excited LPC (VSELP) speech coder, which is a less computationally intense version of code excited LPC (CELP). It uses a number of small codebooks compared to CELP. In CELP, the encoder and decoder share a look-up table, or codebook, of random excitation sequences. The encoder searches the codebook every frame to find the sequence that matches the error signal the closest, and then the index of the best signal is transmitted. It achieves very low bit rates but at the cost of high processing power. IS-95 uses Qualcomm CELP (QCELP), which is based on CELP but has the ability to run at variable rates. There are four data rates in QCELP: full, half, quarter, and eight, and the full rate is about 8 Kbps. The variable rate vocoder is highly beneficial in full-duplex mobile systems, because a typical speaker roughly talks only about 40% of the time. The encoder has 450 frames; 410 at full rate, 15 at half rate, 10 at quarter rate, and 15 at the rate of eight for an average data rate of about 7.5 Kbps, 93.75% of the full rate [4, 5].

In 2.5G and 3G systems some additional vocoders were introduced to further improve the spectral efficiency. The key CDMA2000 (evolution of IS-95) system speech coder is the enhanced variable rate coder (EVRC), which is also based on CELP. The EVRC categorizes speech into full-rate (8.55-Kbps), half-rate (4-Kbps), and eight-rate (0.8-Kbps) frames that are formed every 20 ms. For 2.5G and 3G systems that use a GSM core network, the European Telecommunications Standards Institute (ETSI) recommends the use of GSM-EFR (enhanced full rate) and adaptive multirate (AMR) speech coders. Also, in many GSM networks RPE-LTP has been replaced by better GSM-EFR and AMR vocoders. EFR was developed to improve the poor quality of the GSM-full rate (FR) codec and is based on CELP. AMR, on the other hand, incorporates multiple submodes for use in full-rate or half-rate mode that are determined by the channel quality. The AMR full-rate submode 1 incorporates bit-exact (12.2-Kbps) version of GSM-EFR. Studies have shown that AMR operates better than EFR in heavily loaded networks. The details of these speech coders can be found in [6].

2.3.1.2   Channel Coding

After source encoding, the sequence of binary digits is passed to the channel encoder. The purpose of the channel encoder is to introduce in a controlled manner some redundancy in the binary information sequence, which increases the reliability of the received signal. Channel codes that are used to detect errors are called *error detection codes*, whereas codes that can detect and correct errors are called *error correction codes*. The introduction of redundant bits increases the data rate on the link, hence increasing the bandwidth requirement. The encoder maps the input information sequence into a code sequence for transmission over the channel. The purpose of this mapping is to improve communication efficiency to enabling the system to correct some transmission errors. Forward error correction (FEC) techniques such as Reed-Solomon are used to improve bit error rate performance, whereas automatic repeat request techniques are used to correct errors by requesting retransmission of corrupted data packets.

2G systems use the convolutional code as the error correcting code and the cyclic redundancy code (CRC) as the error detection code for their transmissions. The coding rate R describes the amount of information per encoded bit and is defined as follows:

$$R = k/n \qquad (2.1)$$

where k is the bits per symbol and n is the number of generators.

In addition to convolutional codes, the 3G systems also use turbo codes for high-speed packet data services. The studies and testing have showed that the data throughputs and data capacities are better with turbo coding than with convolutional coding. The forward link throughputs are usually 10% or more better with turbo coding. The details of coding and decoding techniques can be found in [3, 6, 7].

2.3.1.3   Modulation

After channel coding, the binary sequence is passed to the digital modulator, which serves as the interface to the communications channel. The primary purpose of the digital modulator is to map the binary information sequence into signal waveforms.

GSM uses Gaussian minimum shift keying (GMSK), which is a constant envelope modulation scheme in which the phase of the carrier is instantaneously varied by the modulating signal (i.e., the information to transmit). Different variations of phase shift keying (PSK) are used in the 2G and 3G systems. These variations were achieved by changing the phase of the carrier. Differential QPSK (pi/4 DQPSK), binary PSK (BPSK), quadrature PSK (QPSK), offset QPSK (OQPSK), and also quadrature amplitude modulation (QAM) are some of the key modulation and demodulation schemes that are used in mobile communications.

2.3.1.4   The Channel

The communications channel is the physical medium that is used to send the signal from the transmitter to the receiver. In wireless transmission, the channel is usually the atmosphere (free space). As the signal passes through the open air, it is degraded in a variety of ways. The most common form of the signal degradation comes in the form of additive noise, which is generated at the front end of the receiver, where signal amplification is performed. This noise is often called *thermal noise*. In wireless transmission, additional additive disturbances are human-made noise and atmospheric noise picked up by a receiving antenna. Interference is also another form of additive noise, whereas fading is characterized as nonadditive noise. The signal fading occurs due to multipath components, which are time-delayed variations of the same signal arriving at the receiver via different propagation paths [3].

2.3.1.5   Spreading Codes

In addition to the elements described earlier, CDMA systems employ pseudorandom sequence generators to interface with modulators and demodulators. These generators produce a pseudorandom or pseudonoise (PN) binary valued sequence that is used to spread the transmitted signal in frequency at the modulator and to despread the received signal at the demodulator. The multiplication of the input data by a PN sequence having a bit rate much higher than the original data bit rate increases the data rate while adding redundancy to the system [3]. CDMA systems also use Walsh codes in both the forward and reverse links. In the forward link these are used to identify users and in the reverse link for orthogonal modulation [2].

2.4   2G Systems

The 2G systems were introduced to improve voice quality, coverage, and capacity. As mentioned earlier, the four key 2G systems are GSM, USDC, CDMAOne (IS-95), and PDC. This section describes some of the key attributes of these systems [1, 8].

2.4.1   GSM

In 1990, ETSI published the specifications for GSM, which still has, by far, the largest number of subscribers worldwide. GSM is mainly deployed in the 800- and 1900-MHz bands, and there are more than 1 billion GSM users in the world.

    GSM sorts users onto a physical channel using the FDMA multiple-access technique. The channel is divided into frames, during which eight different users share the channel using TDMA. A GSM timeslot or burst period is about $577\,\mu s$, and each subscriber uses the channel for $577\,\mu s$ of every 4.615 ms ($577 \times 8 = 4.615$ ms).

- *Traffic channels:* A traffic channel (TCH) is used to carry speech and data traffic and is defined using a 26-frame multiframe. The length of a 26-frame multiframe is 120 ms. Out of the 26 frames, 24 are used for traffic, 1 is used for the slow associated control channel (SACCH), and 1 is currently is unused. TCHs for the uplink and downlink are separated in time by three burst periods, so that the mobile station does not have to transmit and receive simultaneously.

- *Control channels:* These channels can be accessed both by idle mode and dedicated mode mobiles. The common channels are used by the idle mode mobiles to exchange the signaling information necessary to change to dedicated mode. Mobiles already in dedicated mode monitor the surrounding base stations for handover and other information. The common channels are defined within a 51-frame multiframe, so that dedicated mobiles using the 26-frame multiframe TCH structure can still monitor control channels. The common channels include are broadcast control channel, frequency correction channel, synchronization channel, random-access channel, paging channel, and access grant channel.

2.4.1.1   GSM Architecture

The GSM system consists of three major interconnected systems: the base station subsystem (BSS), the network and switching subsystem (NSS), and the operation support subsystem (OSS). The GSM architecture is very similar to the one shown in Figure 2.1. The other key component is the mobile station itself, which holds the subscriber identity module (SIM) card. A SIM card—one of the great features of GSM—is a memory device that stores the user's identification information.

The MSC is part of the NSS, which also contains the HLR, VLR, and AUC databases.

2.4.2   IS-54 and IS-136

To meet the growing need for cellular services in United States, the TIA adopted the IS-54 standard based on TDMA in 1989. The IS-54 and IS-136 are 2G systems, known as Digital AMPS (D-AMPS). These standards are *dual mode*, which means they support both analog (AMPS) and digital operations.

The IS-54 architecture is quite similar to the generic architecture described in Figure 2.1. The three basic elements are the MS, BS, and NSS. Each base station provides communication to and from mobile terminals in its cell coverage area, and links the mobile station with the switching centers in the NSS. The NSS provides connection to the PSTN and also takes care of the difference in the protocols between IS-54 and the fixed networks. The central unit in the NSS is the MSC, which communicates with the base station through the "A" interface. The MSC performs the handoff procedures of calls from one cell to another as the mobile users move through the service area. The other units, such as the AUC, HLR, and VLR, have the same functions as mentioned earlier.

IS-54 retains the 30-kHz channel spacing of AMPS to facilitate evolution from analog to digital systems. Each frequency channel provides a radio-frequency (RF) bit rate of 48.6 Kbps which is achieved using a pi/4 DQPSK at a 24.3-baud channel rate. This capacity is divided into six timeslots, two of which are assigned to each user so it can support three users simultaneously. It uses the VSELP coder for speech coding. The coder has an output bit rate of 7.95 Kbps and produces a speech frame every 20 ms. Every second, 50 speech frames, each containing 159 bits of speech, are produced for a particular user.

IS-136 added a number of features to the original IS-54 specification, including text messaging, circuit-switched data, and slotted paging channels to support a "sleep" mode in the terminal to conserve mobile battery.

2.4.3   IS-95

IS-95 or CDMAOne is based on CDMA, and it provides more capacity than GSM and IS-54. IS-95 has also been deployed in the 1900-MHz frequency band.

CDMA is a spread-spectrum technology, which means that it spreads the information contained in a particular signal of interest over a much greater bandwidth than the original signal. IS-95 voice call starts with a standard rate of 9.6 Kbps and then it is spread to a transmitted rate of about 1.23 Mbps. Spreading means that digital codes (PN sequences) are applied to the data bits associated with users in a cell. These data bits are transmitted along with the signals of all the other users in that cell. When the signal is received, the codes are removed from the desired signal, separating the users and returning the call to a rate of 9.6 Kbps.

The forward CDMA channel consists of a pilot channel, synchronization channel, a maximum of 7 paging channels, and up to 63 forward traffic channels. The speech coder used in IS-95 is the Qualcomm code excited linear predictive coder (QCELP). And, QPSK and orthogonal QPSK (OQPSK) modulation schemes are used in IS-95 [1].

2.4.4   PDC

The Pacific Digital Cellular (PDC) system is based on TDMA, with three slots multiplexed onto each carrier, similar to IS-54. The channel spacing is 25 kHz with interleaving[1] to facilitate migration from analog to digital. The RF signaling rate is 42 Kbps and the modulation is pi/4 DQPSK. A key feature of PDC is mobile assisted handoff (MAHO), which facilitates the use of small cells for efficient frequency reuse.

Now, it might be worthwhile to see these major 2G standards collectively. Table 2.1 shows a comparison among these standards.

---

1.   The purpose of the interleaving is to add time diversity in a digital communications system without adding any overhead. It spreads the information bits out in time so the bits from a block of source data are not corrupted at the same time. It also provides protection against bursts of errors.

Table 2.1   Key Attributes of 2G Systems

Attributes	GSM	IS-54	IS-95	PDC
Multiple access	TDMA	TDMA	CDMA	TDMA
Duplex mode	FDD	FDD	FDD	FDD
Duplex spacing (MHz)	45	45	45	130 and 48
Carrier spacing (kHz)	200	30	1250	25
Channel per carrier	8 (half rate: 16)	3 (half rate: 6)	64	3 (half rate: 6)
Carrier bit rate (Kbps)	270.8	48.6	1288	42
Frame size (ms)	4.6	40	20	20
Speech coder	RPE-LTP (13 Kbps)	VSELP (7.95 Kbps)	QCELP (var. rate, 8, 4, 2, 1 Kbps)	VSELP (6.7 Kbps)
Channel coding	1/2 rate Convol. + CRC	1/2 rate Convol. + CRC	1/2 (down) and 1/3 (up) rate Convol. + CRC	1/2 rate Convol. + CRC
Modulation	GMSK	pi/4 DQPSK	QPSK, OQPSK	pi/4 DQPSK

2.5   Migration to 2.5G

The growth of Internet and packet data services in the wireline world paved the way for 2.5G and 3G wireless packet data services. The 2.5G systems are built on 2G technologies to provide an intermediate step for operators before they jump on the 3G bandwagon. The 2.5G systems are providing increased data rates and limited data capability. Some of the key 2.5G technologies are discussed next.

2.5.1   HSCSD

High-speed circuit-switched data is an enhancement to the GSM system. It can provide access to four channels simultaneously, thus providing four times the bandwidth (57.6 Kbps) of a standard circuit-switched data transmission of 14.4 Kbps.

2.5.2   GPRS

The general packet data radio system is an evolutionary path for GSM and IS-136. It involves overlaying a packet-based air interface on the existing circuit-switched GSM network. It provides better access to the Internet and other packet data services as compared to GSM. The key advantage (difference) of GPRS is that it is based on packet-switching (PS) technology. PS means that a given user consumes RF resources only when sending or receiving data—unlike 2G circuit-switched technologies where RF resources are utilized even when there are no data to transfer. Contrary to GSM, in GPRS a mobile can have multiple timeslots to achieve higher data rates. Moreover, channel coding schemes are a little bit different than GSM where the most common coding scheme for packet data transfer is Coding Scheme 2 (CS-2). GPRS supports theoretical data rates up to 171.2 Kbps by utilizing all eight 21.4-Kbps channels simultaneously with CS-4. However, in commercial networks data rates in the range of 30 to 80 Kbps are only achievable with good RF planning.

Sometimes all of the timeslots are not available for packet data; few are reserved by operators which further degrades the performance of packet data. The coding schemes and associated data rates are shown in Table 2.2 [9].

The GPRS air interface is similar to GSM and includes some additional logical and control channels and a packet data channel (PDCH) to support data traffic. One key control channel is the packet common control channel (PCCCH), which comprises of number of logical channels. The logical channels of the PCCCH include the following:

- *Packet random access channel (PRACH):* This is only applicable to an uplink and is used by the MS to initiate signaling and packet data transfer.
- *Packet paging channel (PPCH):* This is only applicable in the downlink, and it is used by the network to page an MS prior to packet transfer.
- *Packet access grant channel (PAGCH):* This is only applicable in the downlink, and it is used by the network to assign resources to the MS prior to packet transfer.
- *Packet notification channel (PNCH):* This is used for point-to-multipoint-multicast (PTM-M) notifications to a group of mobile stations.
- *Packet broadcast control channel (PBCCH):* This is not associated with PCCCH, and it is used to broadcast system information to all mobile stations in a cell.

2.5.2.1   GPRS Mobile Classes

There are three classes of GPRS users to effectively utilize both GSM voice and GPRS data services:

- *Class A:* This class supports simultaneous use of voice and data sessions while connecting to both GPRS and GSM networks at the same time. Thus, a Class A user can hold a voice conversation and transfer GPRS data at the same time.
- *Class B:* Class B users can be registered to GPRS and GSM networks at the same time, but using only one or the other at a given time. During a GSM session, GPRS service is suspended, and then resumed automatically after the GSM session has concluded. Most GPRS mobile devices are Class B.
- *Class C:* Class C users are connected to either GPRS or GSM but cannot be attached to both simultaneously. The device must be switched manually between the services.

Table 2.2   GPRS Coding Schemes and Data Rates per Timeslot [9]

Scheme	Modulation	Air Interface Data Rate (Kbps)	Approximate Usable Data Rate (Kbps)
CS-1	GMSK	9.05	6.8
CS-2	GMSK	13.4	10.0
CS-3	GMSK	15.6	11.7
CS-4	GMSK	21.4	16.0

2.5.2.2   GPRS Network

The GPRS architecture shown in Figure 2.3 is very similar to the GSM network. The two key additional elements are the serving GPRS support node (SGSN) and the gateway GPRS support node (GGSN). The SGSN is analogous to MSC/VLR and performs equivalent functions in the packet-switched domain including mobility management, security, and access control functions. The GGSN is the interface between GPRS and external packet data networks (e.g., the Internet). A SGSN can serve multiple BSCs, whereas a BSC can interface with only one SGSN. A SGSN can interface with one or more GGSNs via the Gn interface. The Gn interface uses the GPRS Tunneling Protocol (GTP), which tunnels user data through the IP backbone network between the SGSN and the GGSN.

On activation of GPRS functionality, the MS must attach itself to the GPRS network. This GPRS attachment informs the GPRS network that the MS is available for packet data traffic. The packet data transfer takes place through the establishment of a Packet Data Protocol (PDP) context, which is effectively a data session. A PDP connection represents a virtual connection between an MS and an external packet data network. Every packet data unit (PDU) between the MS and the GGSN is transferred over the appropriate PDP context. The PDP multiplexing is performed by the Subnetwork Dependent Convergence Protocol (SNDCP), which runs between the MS and the SGSN. Normally, such a context is initiated by the MS, but it could also be activated by the network. During PDP context activation, the MS moves from the standby state to the ready state as shown in Figure 2.4 [9, 10].

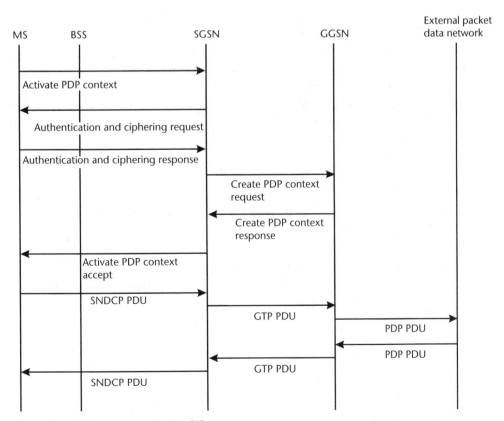

Figure 2.4   PDP context activation [9].

2.6   Migration to 3G

The term 3G was coined by the global mobile industry to indicate the next generation of mobile services. These services include better quality voice, higher capacity,
access to the Internet, and high-speed packet data and multimedia applications. The
international 3G standards have been accepted by the ITU under the name of International Mobile Telecommunications 2000 (IMT-2000). The goal of 3G standards
is to provide users with worldwide coverage via handsets that have the capability to
seamlessly roam among multiple cellular/PCS networks. IMT-2000 is a framework
from the ITU for 3G wireless phone standards throughout the world that deliver
high-speed multimedia data as well as voice. Formerly known as the Future Public
Land Mobile Telecommunications System (FPLMTS), IMT-2000 supports various
technologies that increase data rates such as WCDMA and CDMA2000. The specifications of the different standards for IMT-2000 are being developed by the Third
Generation Partnership Project (3GPP) and 3GPP2 [6, 10].

2.6.1   3GPP

In December 1998 five standard development organizations, ARIB (Japan), ATIS
(USA), ETSI (Europe), TTA (South Korea), and TTC (Japan), agreed to launch
3GPP with the later incorporation of CCSA (China) in May 1999.[2] The aim of this
project is to produce technical specifications for the evolution of GSM systems.

2.6.2   3GPP2

The 3GPP2 was launched to complement the evolution study of non-GSM systems.
It is also a collaborative effort between ARIB, CCSA, Telecommunications Industry
Association—North America (TIA), TTA, and TTC. It has developed global specifi-
cations for the radio transmission technologies (RTTs) supported by the American
National Standards Institute (ANSI). The aim of this project is to produce technical
specifications for the evolution of IS-95 systems.

2.6.3   3G Standards

ITU assigned IMT-2000, to come with a set of standards that can provide the
following:

- Improved system capacity and spectrum efficiency;
- Minimum transmission rate of 144 Kbps (preferably 384 Kbps) in the mobile
  environment (outdoor);
- Minimum transmission rate of 2 Mbps in the fixed (indoor) environment.

---

2.   ARIB, Association of Radio Industries and Businesses; ATIS, Alliance for Telecommunications Industry
Solutions; CCSA, China Communications Standards Association; ETSI, European Telecommunications
Standards Institute; TTA, Telecommunications Technology Association; and TTC, Telecommunications
Technology Committee.

Upon request from the ITU for RTT proposals, different regional standardization bodies submitted their proposals for IMT-2000 in 1998. During the evaluation of the different proposals by the ITU, it turned out that the vision of a global standard with a single radio interface was not realizable for 3G systems. This was due to the different 2G technologies used in the different regions in the world. It would have been impossible to find one technology that could become the evolutionary path for all existing 2G systems. Therefore, a family concept was adopted and agreed on at the end of 1999. The IMT-2000 recommendation highlights five distinct mobile/terrestrial radio interface standards as shown in Figure 2.5:

- *IMT-Direct Spread (IMT-DS):* CDMA Direct Spread is known as WCDMA or WCDMA-FDD. This standard is intended for applications in public macrocell and microcell environments.
- *IMT-Multicarrier (IMT-MC):* IMT-MC refers to CDMA2000 1X and CDMA2000 1xEV (Evolution). 1X means single carrier.
- *IMT-Time Code (IMT-TC):* IMT-TC is a combination of WCDMA-TDD and the TD-SCDMA.
- *IMT Single Carrier (IMT-SC):* IMT-SC corresponds to Universal Wireless Communication 136 (UWC-136) and EDGE (Enhanced Data Rates for GSM Evolution).
- *IMT-Frequency Time (IMT-FT):* IMT-FT is the European Digital Enhanced Cordless Telecommunications (DECT) proposal.

2.6.4 3G Network Evolution

Figure 2.6 shows the evolution of the key 2G technologies to 3G. GSM systems can either migrate to WCDMA or can potentially use the TD-SCDMA radio interface. PDC can evolve to WCDMA, while IS-136 can move to EDGE first and then to WCDMA. 3GPP also recently completed specifications for HSDPA and HSUPA. HSUPA (HSDPA plus HSUPA) is an evolution path for WCDMA. IS-95 systems can take the evolution path toward CDMA2000 and then to its evolution phase. The details of these major radio interfaces are provided in subsequent chapters.

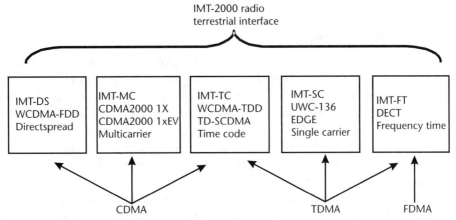

Figure 2.5   IMT-2000 terrestrial radio interfaces.

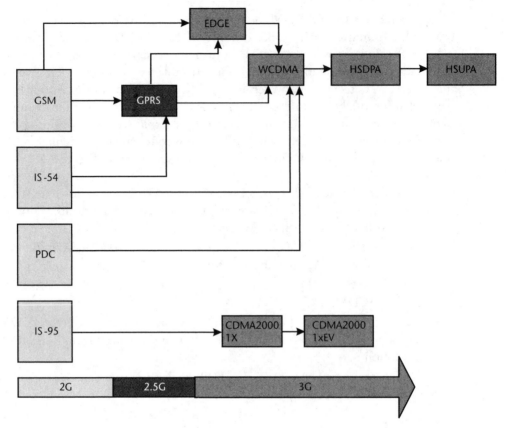

Figure 2.6   2G-to-3G migration.

GSM has the largest subscriber base in the world. It accounts for more than 70% of the world's wireless market and operates in more than 200 countries. The operators are still continuing to use GSM/GPRS networks in low user density areas because of its cost effectiveness over WCDMA. There are more than 1.5 billion users of GSM (including GPRS, EDGE, WCDMA) compared to 250 million plus of CDMA (CDMAOne and CDMA2000), 70 million plus of TDMA, 40 million plus of PDC, and 6 million plus of analog subscribers around the world [11].

2.6.5   Device Evolution

The analog cellular phones were much bigger in size and had lower processing power than the 2G handsets. The use of digital technology and enhancements in digital signal processing (DSP) has led to the development of better quality phones. The analog handsets were mainly voice-centric, whereas the 2G and 2.5G devices also support lower circuit-switched packet data rates. The 3G handsets are steadily evolving from being simplistic voice-centric devices to becoming powerful voice, data, and video communication devices (smart phones) with exceptional computational power.

A *smart phone* is commonly defined as a convergence device that includes a cellular telephone, programmable information-management features, and Internet access. With one such handheld device, a user can access corporate e-mails, personal e-mails, text messages, the Internet, and regular cellular phone functions such as a phone book, internal storage (memory), calendar, and so forth. Smart phones also allow users to play 3D games, listen to MP3s and FM radio, and record and play back videos. Many newer smart devices offer Bluetooth connectivity that wirelessly connects to nearby devices, such as headsets, PCs for synchronization, and other smart phones.

2.7 Summary

This chapter highlighted the evolution of mobile communications that began with analog technology in the 1970s. It illustrated the key concepts, the process of information flow and briefly described the key 1G and 2G technologies. It also presented the evolutionary path (both network and devices) toward 3G.

References

[1] Rappaport, S. T., *Wireless Communications Principles and Practice*, Upper Saddle River, NJ: Prentice Hall, 1996.

[2] Yang, C. S., *3G CDMA2000 Wireless System Engineering*, Norwood, MA: Artech House, 2004.

[3] Proakis, G. J., and M. Salehi, *Communications Systems Engineering*, Englewood Cliffs, NJ: Prentice Hall, 1994.

[4] William, L., *Mobile Cellular Telecommunications: Analog and Digital Systems*, New York: McGraw-Hill, 1995.

[5] Redl, H. S., K. M. Weber, and W. M. Oliphant, *An Introduction to GSM*, Norwood, MA: Artech House, 1995.

[6] Garg, V. K., *Wireless Network Evolution*, Upper Saddle River, NJ: Prentice Hall, 2002.

[7] Asif, S., "Data Coding Techniques in 3G1x RTT Systems," *Proc. 2002 World Wireless Congress*, San Francisco, CA, May 28–31, 2002, pp. 571–575.

[8] Padghett, E. J., G. C. Gunther, Hattori, "Overview of Wireless Personal Communications," *IEEE Communications Magazine*, Vol. 33, No. 1, January 1995, pp. 28–41.

[9] Smith, C., and J. Meyer, *3G Wireless with WiMax and Wi-Fi: 802.16 and 802.11*, New York: McGraw-Hill, 2004.

[10] Willie, W. L., *Broadband Wireless Mobile*, New York: John Wiley & Sons, 2002.

[11] GSM Association, http://www.gsmworld.com.

CHAPTER 3

Introduction to CDMA2000

3.1 Introduction

The migration path from second generation CDMAOne or IS-95 as defined by IMT-2000 is CDMA2000. The CDMA2000 family of standards defines a spread-spectrum radio interface that uses the CDMA technology to meet the requirements for 3G wireless communication systems [1, 2]. The term 3G is associated with higher voice capacity and high data rates to support advanced services and applications. CDMA2000 includes both IS-2000-based and IS-856-based radio access technologies. IS-2000 systems were first standardized and commercialized before IS-856 networks. The IS-2000 standard is focused on improving traditional voice and providing packet data services, whereas IS-856 is designed for higher packet data rates and Voice over Internet Protocol (VoIP) services. We will make an attempt to address both sets of technologies in this chapter.

Section 3.2 explains IS-2000 technology including its physical, link, and upper layers. Section 3.3 illustrates the two migration paths of CDMA2000 1X technology: CDMA2000 1xEV-DO (evolution data optimized or IS-856) and CDMA2000 1xEV-DV (evolution data and voice; based on IS-2000). Section 3.4 lists the enhancements of CDMA2000, Section 3.5 describes IS-2000- and IS-856-based network architectures, Section 3.6 provides a futuristic view of CDMA2000 technologies, and Section 3.7 compares the different technologies. Section 3.8 provides a short summary.

3.2 CDMA2000 1X (IS-2000)

CDMA2000 is an ITU-approved 3G wireless communications standard. CDMA2000 1X (IS-2000 or 3G1x) was first commercialized in October 2000 in South Korea and since then, more than 70 operators on six continents have launched CDMA2000 services. The networks and devices supporting these services are backward compatible with those based on 2G CDMAOne. The backward compatibility helps in preserving operator spectrum and equipment investments, while increasing voice capacity and providing effective data capability.

3.2.1 IS-2000 Carrier Spacing, Spreading Rate, and Radio Configuration

The IS-2000 technology is comprised of multiple radio transmission technologies (RTTs) where 1xRTT utilizes a single 1.25-MHz carrier (same as CDMAOne systems). The subsequent RTTs may support multiples of the 1.25-MHz bandwidth;

for example, 3xRTT will require a 3.75-MHz band in both the forward and reverse directions. CDMA2000 networks have been primarily deployed in the 800-MHz and PCS (1900-MHz) bands. In 800 MHz, the A and B blocks have 11 and 9 MHz of bandwidth. In PCS, the spectrum is allocated either in 5-MHz blocks (D, E, and F) or 15-MHz blocks (A and C). Few additional frequency bands have been defined in the standard. Due to the multitude of frequency bands, the CDMA2000 operators have signed roaming agreements with other service providers to enable services to their customers globally.

IS-2000 RTT technology specifies two modes of spreading in the downlink: direct spread (DS) and multicarrier (MC). The modes are based on how the signals are spread over the entire bandwidth. The MC option allows spreading of the signal onto three separate 1.25-MHz carriers, at the rate of 1.2288 megachips per second (Mcps) per carrier for a total chiprate of 3.6864 Mcps across the 3.75-MHz bandwidth (Figure 3.1). The DS option on the other hand spreads the signal over the entire 3.75-MHz band, at a chiprate of 3.6864 Mcps (Figure 3.1). In the uplink IS-2000 specifies spreading the CDMA signal across the entire bandwidth of three 1.25-MHz carriers, using a chiprate of 3.6864Mcps (i.e., DS). Spreading rate (SR) is a new parameter introduced in IS-2000. SR is the pseudonoise chip rate of the system defined as a multiple of 1.2288 Mcps. Two different spreading rates are covered by IS-2000:

- *Spreading rate 1 (SR1):* The 1.2288-Mcps chiprate-based system uses a DS single carrier. SR1 is used for both CDMAOne and 3G1x systems.
- *Spreading rate 3 (SR3):* A 3.6864-Mcps chiprate-based system uses three 1.2288-Mcps carriers or a single 3.6864-Mcps DS carrier on the forward

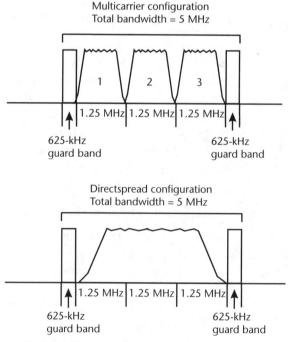

Figure 3.1   Directspread and multicarrier configurations.

channel and a 3.6864-Mcps DS carrier on the reverse channel. Spreading rate 3 is specified to be used with the 5-MHz systems.

CDMAOne traffic channel data rates are grouped into groups called *rate sets*. Rate set 1 (RS1) and rate set 2 (RS2) are the rate sets supported by CDMAOne and used with 8- and 13-Kbps vocoders, respectively. The 8-Kbps vocoders, EVRCs, and 13-Kbps vocoders have been used for 3G1x systems but new radio configurations (RCs) are defined to take into account the different spreading rates and modulation. RCs define the data rates that will be used in IS-2000 networks. Nine RCs have been classified for IS-2000 forward link and six for reverse link. RC1 and RC2 are the same as RS1 and RS2 in CDMAOne systems. RS1 includes data rates of 1,500, 2,700, 4,800, and 9,600 bps, whereas RS2 includes 1,800, 3,600, 7,200, and 14,400 bps. Beside other factors, the data calls are heavily dependent on the type of RCs that are enabled by the service provider and availability of the power in the BTS. The details of RCs can be found in the IS-2000 standard [3, 4].

3.2.2   Background

The first step in migration from 2G (CDMAOne) technology is CDMA2000 1X, which provides twice the voice capacity and higher data rates. The most fundamental change comes from the support for data services via wireless devices. CDMA2000 1X can support a 307.2-Kbps peak data rate in both forward and reverse directions. Beyond CDMA2000 1X, this technology has evolved into two separate paths: 1xEV-DO (IS-856) and 1xEV-DV (IS-2000 evolution) as illustrated in Figure 3.2. The original idea of having a multicarrier downlink was envisioned to meet the IMT-2000 requirements, with 3X as the minimum compliant level. The 3X idea did not get much industry support because of the technical complexities both at the base stations and handsets, uncertainty around the need for data services, and cost of migration from 2G. As the penetration of data increases, the 3X has been supplanted by standards such as 1xEV-DO and 1xEV-DV.[1] Our focus is mainly only on single carriers, so SR3 is not addressed in this chapter.

3.2.3   CDMA2000 1X Layering Structure

Figure 3.3 shows the CDMA2000 1X (IS-2000) layering structure, which illustrates the breakdown in terms of the upper layers, the link layer, and the physical layer. At

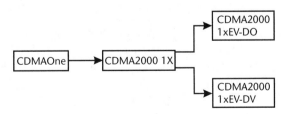

Figure 3.2   CDMA2000 evolution.

---

1.   In May 2006, the 3GPP2 completed the multicarrier EV-DO (EVDO Release B) standard, which specifies the support for 3 concatenated carriers (3X), at a minimum.

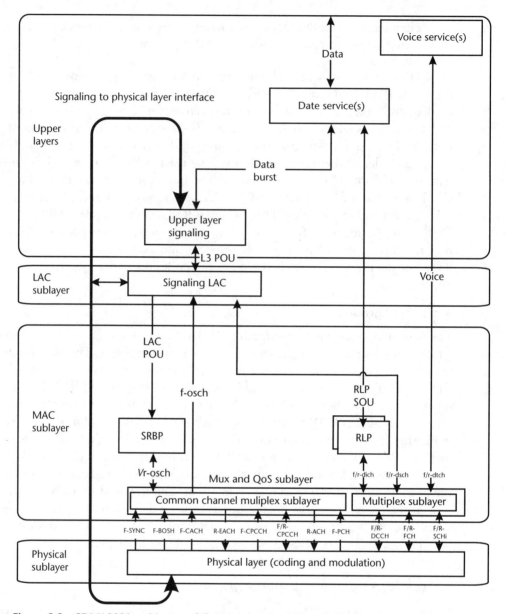

Figure 3.3 CDMA2000 architecture [2]. Reproduced under written permission from Telecommunications Industry Association.

the most basic level, IS-2000 provides protocols and services that correspond to the bottom two layers (low layers): the physical layer and link layer. The Link Layer is further subdivided into the link access control (LAC) sublayer and the medium access control (MAC) sublayer [1].

3.2.3.1 Physical Layer

CDMA2000 1X physical layer (air interface) is one of the major critical elements of the technology. The physical layer provides channels, coding, and modulation schemes to transfer data between the mobile device and the network. To support voice and data services, CDMA2000 1X specifies a number of new channels, in addition to the existing channels currently in IS-95 (CDMAOne). The channels specified for both IS-2000 Releases 0 and A are discussed next.

CDMA2000 1X Channels
CDMA2000 1X defines two categories of channels: logical and physical. Logical channels carry control and user data information whereas, the physical channels map logical channel data to a particular radio resource.

Logical Channels.   A logical channel name consists of three lowercase letters followed by "ch" (channel). A hyphen is used after the first letter (Table 3.1). The combinations of f/r-dtch, f/r-ctch, f/r-dmch, f/r-cmch, dsch, and csch are possible to support various functions. These combinations support data transfer, synchronization, access, and dedicated signaling. The different logical channels are mapped to physical channels. Some logical channels can be assigned to several different physical channels, and this is called the physical layer dependent convergence function (PLDCF). Other logical channels provide data transport without being dependent on the underlying physical layer, and this is called physical layer independent convergence function (PLICF).

Physical Channels.   Figure 3.4 illustrates a current view of the CDMA2000 1X physical channel structures on the forward and reverse links, which are included in the initial (IS-2000 Release 0) and subsequent release (IS-2000 Release A) of CDMA2000 1X [3, 4].

Forward Link Channels' Description: IS-2000 Release 0.   Forward link channels including traffic and common carry information from the base station to the mobile station. The common channels are pilot, page, and sync, and the traffic channels are fundamental and supplemental:

- *Forward pilot channel (F-PICH):* An unmodulated, direct-sequence, spread-spectrum signal transmitted continuously by a CDMA BS. A pilot channel provides a phase reference for coherent demodulation and provides a means for signal strength comparisons between base stations for determining when a handoff is required.
- *Forward synchronization channel (F-SYNC):* A code channel on the forward CDMA channel, which transports the synchronization messages to the mobile station. Synchronization is used to bootstrap the mobile's long code state, frame timing, and configuration of the paging channel.

Table 3.1   Naming Conventions for Logical Channels [1]

First Letter	Second Letter	Third Letter
f = forward	d = dedicated	t = traffic
r = reverse	c = common	m = Media access control
		s = signaling

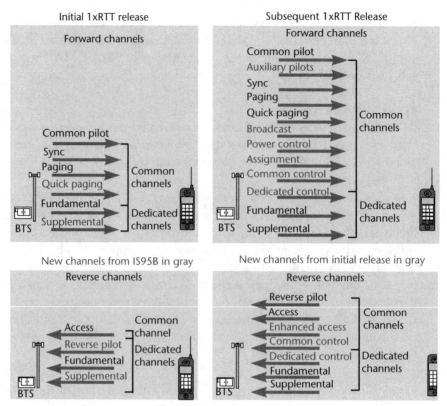

Figure 3.4 CDMA2000 physical channels.

- *Forward paging channel (F-PCH):* Carries overhead messages, pages, acknowledgments, channel assignments, and status requests. Shared secret data updates from the BS to the MS as described later in this chapter.
- *Forward quick paging channel (F-QPCH):* Used by a BS when it needs to contact the MS in slotted mode.[2]
- Each *forward traffic channel (F-TCH)* contains one fundamental channel and zero or one supplemental channel.
  - *Forward fundamental channel (F-FCH):* Transmitted at different rates (depending on the RC) negotiated at the beginning of the call. Each F-FCH is transmitted on a different orthogonal code channel and supports frames of 20 ms.
  - *Forward supplemental channel (F-SCH):* Used for high-speed packet data calls and also operates at prenegotiated variable rates. Except for RC1 and RC2, all other RCs support F-SCH.

Reverse Link Channels' Description: IS-2000 Release 0. Reverse link channels including traffic and common channels carry information from the MS to the BS. The common channels are access and pilot, and the traffic channels are fundamental

2. *Slotted mode:* An operation mode of the moblie station in which the mobile station monitors only selected slots on PCH. This mode saves mobile battery life.

and supplemental. The structure for the ACCESS CHANNEL (R-ACH) is the same for SR1. This means that the access channel used for 3G1x is the same as the existing access channel used for IS-95.

- *Reverse pilot channel:* A direct-sequence, spread-spectrum signal transmitted for the duration of the call from the mobile station. The reverse link pilot channel provides a phase reference for coherent demodulation and fast forward power control, 800 times per second.
- Each *reverse traffic channel (R-TCH)* contains one fundamental channel and zero to one supplemental channel:
    - *Reverse fundamental channel (R-FCH):* Transmitted at different rates. The rates supported for the R-FCH are 1.5, 2.7, 4.8, and 9.6 Kbps for RC3 and RC5 and 1.8, 3.6, 7.2, and 14.4 Kbps for RC4 and RC6.
    - *Reverse supplemental channel (R-SCH):* Used for data calls and can also operate at different prenegotiated rates.

Forward Link Channels' Description: IS-2000 Release A.   The following subsections describe only the new channel structures from the initial 3G1x release in the forward link.

- *Auxiliary pilot channel:* Either common or dedicated to a single user. It is an unmodulated, direct-sequence spread-spectrum signal transmitted continuously by a BS. An auxiliary pilot channel is required for forward link spot beam and antenna beam forming applications, and provides a phase reference for coherent demodulation of forward link CDMA channels associated with the auxiliary pilot.
- *Broadcast channel (F-BCCH):* A coded channel in a forward CDMA channel used for transmission of control information and pages from a BS to a MS. It operates with the code rates of 1/4 and 1/2 and data rates of 4.8 to 19.2 Kbps.
- *Common power control channel (F-CPCH):* A forward common channel that transmits power control bits (i.e., common power control subchannels) to multiple mobile stations.
- *Common assignment channel (F-CACH):* A forward common channel used by the BS to acknowledge a mobile station accessing the Enhanced Access Channel.
- *Forward Common Control Channel (F-CCCH):* Used by the BS to transmit mobile station-specific messages.
- *Forward Dedicated Control Channel (F-DCCH):* Used to carry user data as well as signaling and control information while the call is in progress. DCCH is necessary to support supplemental channel and simultaneous voice and data.

Reverse Link Channels' Description: IS-2000 Release A.   The following subsections describe only the new channel structures from the initial 3G 1x release in the reverse link:

- *Enhanced access channel (R-EACH):* A reverse channel used by the mobile for communicating to the BS. It is used for transmission of short messages such as signaling, MAC messages, response to pages and call originations, with higher rates, lower latency (slot duration ACK timeout and so on), all rates and frame sizes. Also, it can be used to transmit moderate-sized data packets.
- *Reverse common control channel (R-CCCH):* A portion of a reverse CDMA channel used for the transmission of digital control information from one or more MSs to a BS.
- *Reverse dedicated control channel (R-DCCH):* Used for transmission of higher level data and control information from a MS to a BS.

3.2.3.2 Link Layer

The link layer is further subdivided into MAC and LAC sublayers as shown in Figure 3.5.

MAC Sublayer

MAC is an entity that controls the access to and from the upper layer signaling, data services, and voice services to physical layer resources. MAC also coordinates and reserves the air interface resources required to maintain a given quality of services. The CDMA2000 MAC sublayer provides best effort delivery of data over the radio link with a Radio Link Protocol (RLP), multiplexing, and quality of service (QoS) control via prioritization of requests [5].

One of the main functional entities of the MAC layer is the multiplexing and QoS sublayer. The multiplexing and QoS sublayer is subdivided into a multiplex sublayer and a common channel multiplex sublayer for transmission and reception of dedicated and common information and signaling, respectively. The multiplex sublayer has both transmitting and receiving functions, as depicted in Figure 3.4. The multiplex sublayer multiplexes one or more data blocks into a multiplex

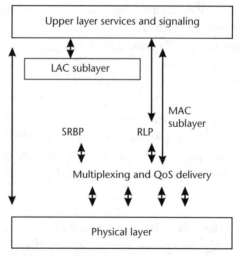

Figure 3.5 MAC and LAC sublayers in CDMA2000 layer structure [6].

sublayer packet data unit (MuxPDU). The multiplex sublayer transmitting function combines information from various sources (e.g., upper layer signaling, voice, and data services) and forms physical layer service data units (SDUs) for transmission. The multiplex sublayer receiving function separates the information contained in physical layer SDUs, and directs the information to the correct entity (e.g., upper layer signaling, voice, or data services). The common part of the multiplex sublayer establishes the common physical channels. Also, a Signaling Radio Burst Protocol (SRBP) enables a connectionless service for signaling messages between the LAC and multiplexing and QoS sublayers.

LAC Sublayer

Link access control is an entity that provides reliable and unreliable delivery of information across the air interface between the MS and the BS. It also controls errors and sets up, maintains, and releases logical connections.

The LAC sublayer implements a data link protocol to deliver signaling messages generated by layer 3. SDUs are passed between layer 3 and the LAC sublayer and they are encapsulated into LAC PDUs. These PDUs are subject to segmentation and reassembly and are transferred to the MAC sublayer. SDUs and PDUs are processed and transferred along functional paths, the logical channels, without the need for the upper layers to be aware of the radio characteristics of the physical channels. As a generated or received data unit traverses the LAC protocol stack, it can be processed by the following five sublayers of LAC (Figure 3.6) [7]:

- *Authentication sublayer:* This sublayer is responsible for execution- and authentication-related functions. The authentication sublayer is only used on the reverse link common signaling channel (r-csch).

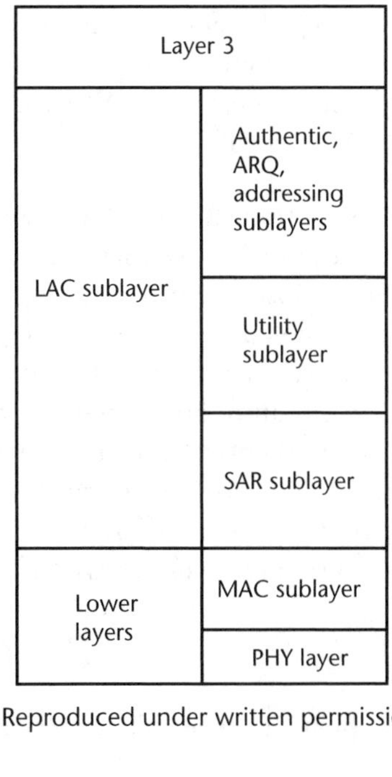

Figure 3.6   LAC sublayers [7]. Reproduced under written permission from Telecommunications Industry Association.

- *ARQ sublayer:* This layer provides the functionality to support reliable exchange of upper layer SDUs, detecting and discarding duplicate packets. The IS-2000 ARQ sublayer uses a selective repeat request approach to retransmit only the lost units. The receiving end is required to store the correctly received packet data units. The ARQ is used on both forward and reverse logical channels including f/r-csch, f/r-dsch, and f/r-dmch. The ARQ sublayer for each logical channel is split between the transmitter and the receiver side and provides both assured delivery and unassured delivery services to layer 3. In assured delivery service the PDUs are repeatedly sent at fixed intervals until an acknowledgment is received or a specific number of retransmissions is reached, in which case the logical channel is dropped. In unassured delivery service PDUs are not acknowledged, and there is no guarantee of reception.

- *Addressing sublayer:* This sublayer is responsible for managing the address-related fields in the messages. The addressing sublayer adds address fields to the SDU and removes them from PDUs in transmit and receive directions, respectively. This sublayer supports International Mobile Subscriber Identity (IMSI), Electronic Serial Number (ESN), Temporary Mobile Subscriber Identity (TMSI), a combination of IMSI and ESN, and a combination of IMSI_Short and ESN types of addresses. This sublayer is used on the forward and reverse common signaling channels.

- *Segmentation and reassembly (SAR):* This sublayer also has both transmitting and receiving functions. On the transmitting side it segments the PDU, and on the reverse side it reassembles to form the original PDU. SAR is responsible for adding the length field as a prefix and the CRC field at the end of the packet. A PDU can be divided into many blocks, and this limit is not defined in the IS-2000 standard.

- *Utility sublayer:* This sublayer is used in both forward and reverse directions on the common signaling channel, dedicated signaling channel, and dedicated MAC channel in both directions. The utility sublayer is responsible for converting/recovering the SDU_TAG of the SDU to/from the MSG_TYPE of PDU. PDUs with unknown MSG_TYPE are discarded. Encryption parameters are included on an as-needed basis.

The preceding LAC description showed that each LAC sublayer has a well-defined function. A common feature of all of these sublayers is that they add fields in the transmit direction and remove them in the reverse direction. Also, each sublayer does some processing on the payload unit of the packet. At the same time, it is not necessary to have all five sublayers supported on every channel. For example, if a service provider does not require an authentication procedure in the network, the operator can disable the authentication sublayer functions. The IS-2000 standards do not specify the interfaces between different LAC sublayers. These interfaces may be proprietary and may be handled differently at the BS and MS. The implementation will also vary on these interfaces across different vendors.

3.2.3.3 Upper Layers

Layer 3 (upper layers) provides three main services [8]:

1. *Voice service,* including mobile-to-mobile calls and mobile-to-landline calls.
2. *Circuit and packet data services,* including dial-up access, asynchronous fax services, IP services, and so forth.
3. *Signaling,* including services that control different aspects of the wireless operation.

Layer 3 signaling for IS-2000 as shown in Figure 3.7 is divided into two planes, data and control. The data plane is used for signaling protocol, and the control plane is used for supervision of the signaling protocol according to the functional requirements. The interface between layer 3 and layer 2 is a service access point (SAP), at which SDUs and control information in the form of message control and status blocks (MCSBs) are interchanged. SAPs are defined only for the data plane, and no SAPs are defined for communication through the control plane. Layer 3 generates layer 3 PDUs and passes these PDUs to lower layers, where proper encapsulation into lower layer PDUs is performed. On the receiving end, lower layer PDUs are decapsulated, and the resulting SDUs are sent from the lower layers to layer 3 for processing.

Upper Layer—Reverse Link

Layer 3 assists in the security, authentication, call processing, handoffs, tiered services, and hashing aspects of the MSs. This section briefly highlights these aspects.

- *Security and identification:* Mobile stations are mainly identified by their IMSI. The IMSI consists of up to 15 numerical characters (0–9). The first three digits of the IMSI are the Mobile Country Code (MCC), and the remaining digits are the National Mobile Station Identity (NMSI), consisting of the Mobile Network Code (MNC) and the Mobile Station Identification Number (MSIN). For example, the mobile could have an IMSI of 250-09-301-123-1234, where 250 is the MCC, 09 is used for the identification of mobile operator, and the rest is the subscriber's mobile number.

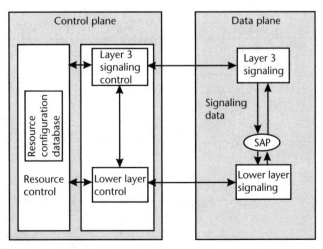

Figure 3.7 IS-2000 signaling architecture [9]. Reproduced under written permission from Telecommunications Industry Association.

- *Authentication:* IMSI_M (MIN-based IMSI) and IMSI_T (True IMSI) are used for the authentication process: IMSI_M will be used if it is available; otherwise, IMSI_T will be used. A successful outcome of the authentication process occurs only when it can be demonstrated that the MS and the BS possess identical sets of shared secret data (SSD). The SSD is a 128-bit quantity that is stored in semipermanent memory in the MS and is readily available to the BS (Figure 3.8). The SSD is divided into two distinct 64-bit subsets. Each subset is used to support a different process: SSD_A is used to support the authentication procedures, and SSD_B is used to support voice privacy and message encryption. The SSD is not accessible to the user.

- *Privacy and encryption:* Voice privacy is provided in the CDMA system by means of the private long code mask (PLCM) used for PN spreading. Voice privacy is provided on the traffic channels only. All calls are initiated using the PLCM for PN spreading, but the MS user may request voice privacy during or after call setup. In an effort to enhance the authentication process and to protect sensitive subscriber information, a method is provided to encrypt selected f/r-dsch or f/r-csch layer 3 signaling PDUs. Before going through the encryption algorithm, the sender of the message can append an 8-bit CRC to the end of the layer 3 PDU. The generator polynomial for the 8-bit CRC field is $g(x) = x^8 + x^7 + x^4 + x^3 + x + 1$. The CRC computation is shown in Figure 3.9.

- *Call processing:* Mobile Station layer 3 processing consists of the following states:

  - *Mobile station initialization state:* The MS selects and acquires a system.
  - *Mobile station idle system:* The MS monitors messages on the f-csch.
  - *System access state:* The MS sends messages to the BS on the r-csch.
  - *Mobile station control on the traffic channel state:* The MS communicates with the BS using the f-dsch and r-dsch.

After the mobile is powered up, it enters the system determination substate of the MS initialization state with a power-up indication. It then enters the pilot

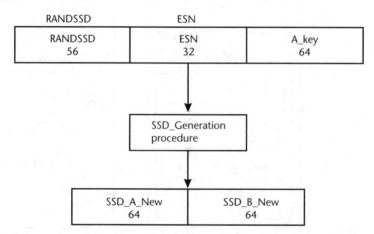

Figure 3.8 Computation of shared secret data [8]. Reproduced under written permission from Telecommunications Industry Association.

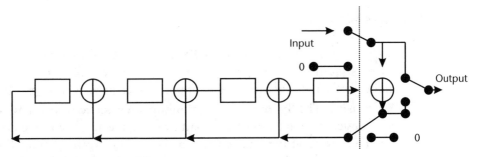

Figure 3.9   CRC generation [6].

channel and sync channel acquisition states and then finally enters the MS idle state. If the call is being placed, it then enters the system access state. When the access attempts are successful it enters the traffic channel state.

- *Handoffs:* The MS supports the following four handoff procedures while in the MS control on the traffic channel state:
  - *Soft handoff:* A handoff in which the MS begins communications with a new BS without interrupting communications with the old BS. These types can only be used between CDMA channels having same frequency assignments.
  - *Softer handoff:* A handoff in which the MS begins communications with a new sector of the same BS without interrupting communications with the old sector of the BS. These types can only be used between CDMA channels having the same frequency assignments.
  - *CDMA-to-CDMA hard handoff:* A handoff in which the MS is transitioned between disjoint sets of BSs, different band classes, different frequency assignments, or different frame offsets.
  - *CDMA-to-analog handoff:* A handoff in which the MS is directed from a CDMA traffic channel to an analog voice channel.
- *Hash function:* The BS or MS is presented with a number of resources and must choose which one to use. For example, if there are seven active paging channels, the BS must decide which of these seven channels to use when paging a particular mobile user. The choice should be uniform and reproducible. One method of making a uniform choice among a number of resources is by using a *hash function*. The hash function allows this uniform form distribution through a method that is reproducible at both MSs and BSs using as inputs (1) the number of resources and (2) a parameter (hash key) that is known by both sides.
- *CDMA tiered services:* The MS can support tiered services based on user zones while in mobile station idle state and in the MS control on the traffic channel state. Tiered services provide custom services and special features to the user based on the MS location.

Upper Layer—Forward Link

- *Security and authentication:* If the BS supports MS authentication, it sends and receives authentication and encryption control messages, performs the authentication calculations, and supports voice privacy using the PLCM.
- *Supervision:* The BS continuously monitors each active access channel. The BS provides control in cases of overload by using the access parameters message. The BS continuously monitors each active reverse traffic channel (R-TCH) to determine if the call is active. If the BS detects that the call is no longer active, the BS announces loss of R-TCH continuity.
- *Registration:* Registration is the process by which a MS notifies the BS of its location, status, identification, slot cycle, protocol revision number, and other characteristics. The CDMA system supports 10 different forms of registration to address these characteristics. The 10 registration forms can be found in the IS-2000 standard.
- *Call processing:* Base station processing consists of the following types of processing:
    - *Pilot and sync channel processing:* During this stage, the BS transmits the pilot channel and sync channel that the mobile uses to acquire and synchronize to the CDMA system while the MS is in the MS initialization state.
    - *Paging channel processing:* During this stage, the BS transmits the paging channel that the MS monitors to receive messages while the MS is in the MS idle state and the system access state.
    - *Access channel processing:* During this stage, the BS monitors the access channel to receive messages that the MS sends while the MS is in the system access state.
    - *Traffic channel processing:* During this stage, the BS uses the forward and reverse traffic channels to communicate with the MS while the MS is in the MS control on the traffic channel state.
- *Handoff and tiered services:* The BS supports different types of handoffs and tiered services as described in the earlier section.

Call Processing

Call processing is the complete process of routing, originating, and terminating wireless voice and data calls. In CDMA2000 systems several types of call and data processing take place. Depending on whether the service being requested or offered is circuit switched (CS) or packet switched (PS), the call processing methods are a little bit different. Irrespective of CS or PS services, the system will perform handoffs and maintain power control. Though there are differences in how call processing takes place for CS and PS services, the fundamental concepts of how neighbors (sectors) are promoted and demoted is still the same.

Numerous call flows occur in CDMA2000 systems, Figures 3.10 and 3.11 explain the two key ones. Figure 3.10 explains the CDMA call origination process, which includes traffic channel initialization, conversation, and release (not shown in figure).

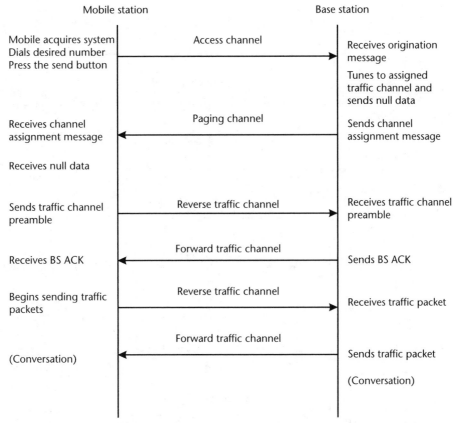

Figure 3.10   Flow diagram for CDMA call origination [9]. Reproduced with permission of The McGraw-Hill Companies.

Figure 3.11 depicts the CDMA call termination call flow, which consists of traffic channel initialization, waiting for order (MS receives alert info from BS), waiting for mobile answer, conversation, and release (not shown in figure). The details about call processing can be found in [4, 8, 9].

3.2.4   CDMA Capacity

To a large degree, the performance of CDMA systems is interference limited. This means that the capacity, which is defined as the total number of simultaneous users the system can support, is limited by the amount of interference present in the frequency band. The capacity of a CDMA cell depends on many different factors including receiver demodulation, power control accuracy, cochannel interference, interference from other users in the same cell and in neighboring cells, and so forth.

In digital communications, the link metric is E_b/N_0 or energy per bit per noise power density. The CDMA capacity M (total number of users in the band) is given by the following formula and the details can be found in [10]:

$$M = \left[(W/R) \times \lambda\right] / \left[E_b/N_o \times (1+\eta) \times v\right] + 1 \tag{3.1}$$

where:

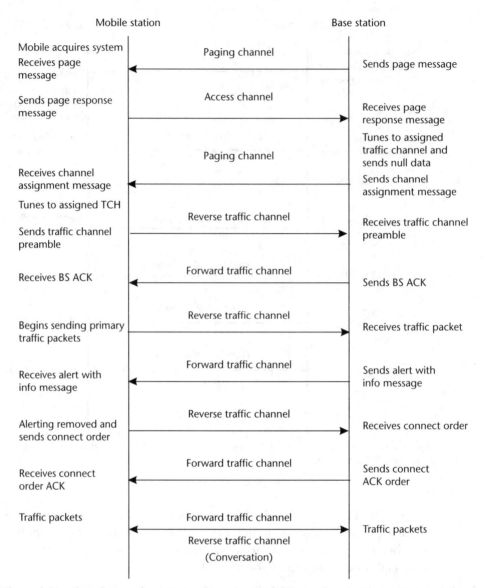

Figure 3.11 Flow diagram for CDMA call termination [9]. Reproduced with permission of The McGraw-Hill Companies.

W/R = processing gain = bandwidth/data rate;

λ = sectorization gain (2.5 for a three-sector cell);

v = voice activity factor (0.4–0.5);

η = loading factor adjacent cell interference (0.5–0.6).

Capacity could be hard, as in the case of analog, or soft, as in the case of CDMA, which is defined in (3.1). Soft capacity is limited by the amount of interference in the air interface; less interference means more channels and more users. Hard capacity is limited by the amount of hardware, output power, and so forth. For an analog system, traffic planning involves examining the hard capacity of each cell and comparing it with the projected traffic load for that cell. If the projected traffic load is greater than the hard capacity or soft capacity of that cell, then an additional cell needs to be built in the vicinity.

There are two major types of calls: voice calls and data calls. Voice capacity is normally measured in Erlangs. An Erlang is defined as the average number of simultaneous calls. In terms of packet data, the capacity is the product of total number of users and their corresponding average data rates per sector.

Current CDMA systems are mainly designed to support voice and low-bandwidth data traffic. These are not designed to handle high-bandwidth video streaming (TV). The networks are designed to be *unicast*, which means that signals are transmitted between a single sender and a single receiver. These unicast networks are not designed to cover big media events such as a breaking news story or big sporting events.

3.3   Evolution of CDMA2000 1X

CDMA2000 1X has evolved into two separate paths: CDMA2000 1xEV-DO and CDMA2000 1xEV-DV. The 1X refers to a single carrier and 1xEV represents its evolution path as shown in Figure 3.12.

The 1xEV-DO (IS-856) has two revisions, Revision 0 and Revision A. 1xEV-DO Revision 0 mainly improves the forward link data rates of CDMA2000 1X and has been deployed in some countries around the world. Revision A, which was standardized in early 2004, not only increases data rates in the forward link but also adds a number of enhancements in the reverse link. Neither of the two revisions supports circuit-switched voice but Revision A has the capability to support VoIP [11, 12].

The 1xEV-DV (IS-2000 Release D) supports high data rates in both forward and reverse links and it supports circuit-switched voice. IS-2000 Release C concentrates on the forward link improvements, and those are also included in the IS-2000 Release D, which also added enhancements in the reverse link. Release D was also standardized in early 2004, as was IS-856-A, and both of these technologies have similar capabilities as described in this and the following sections. The IS-856 and IS-2000 revisions/releases are not backward compatible to each other. On the other 1xEV-DV is backward compatible to CDMA2000 1X standard, and 1xEV-DO Revision 0 is backward compatible to 1xEV-DO Revision A [13, 14].

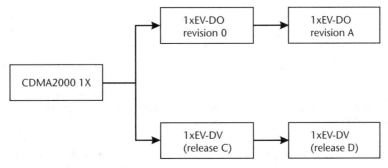

Figure 3.12   Evolution of CDMA2000 1X technology.

3.3.1 CDMA2000 1xEV-DO

1xEV-DO Revision A technology has been standardized by 3GPP2, and it supports data with peak data rates of 3.072 Mbps in the downlink and 1.8 Mbps in the uplink. It supports VoIP and QoS and provides battery life improvements. 1xEVDO-A is interoperable with 1xRTT voice and packet data and backward compatible with Release 0 of IS-856 (1xEVDO), bringing more features and performance via higher data rates on the forward and reverse link plus quality of service for real-time application support. It operates in the PCS frequency band and other bands specified in the standard.1xEV-DO-A will require a separate carrier (frequency) to operate, and it cannot coexist on the same carrier with CDMA2000 1X. Though the focus of this document is on the 1xEV-DO Revision A[3] physical layer, it also highlights the description of the remaining layers and protocols.

3.3.1.1 1xEV-DO Architecture Model

The 1xEV-DO architecture model consists of the following functional units, as shown in Figure 3.13:

- *Access terminal (AT):* AT is a device that provides data connectivity to a user. An access terminal may be connected to a computing device, such as a laptop, or may be a self-contained data device, such as a personal digital assistant.
- *Access network (AN):* AN equipment provides data connectivity between a packet data network (typically the Internet) and the access terminal.
- *Sector:* The part of the access network that provides the land-side modem.

3.3.1.2 Layering Architecture

The air interface has been layered, with an interface defined for each layer and for the protocol within each layer. This allows for future modifications to a layer or to a protocol to be isolated. Figure 3.14 describes the layering architecture for the air interface. Each of the seven layers consists of one or more protocols that perform the layer's functionality. Each of these protocols can be individually negotiated.

A brief description of each layer follows:

- *Application layer:* Provides multiple applications and handles transport of protocol messages and user data;

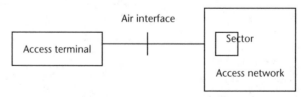

Figure 3.13 Architecture reference model [11, 12]. Reproduced under written permission from Telecommunications Industry Association.

---

3.   1xEV-DO Revision 0 (EVDO-0) physical layer is a subset of 1xEV-DO Revision A (EVDO-A) physical layer.

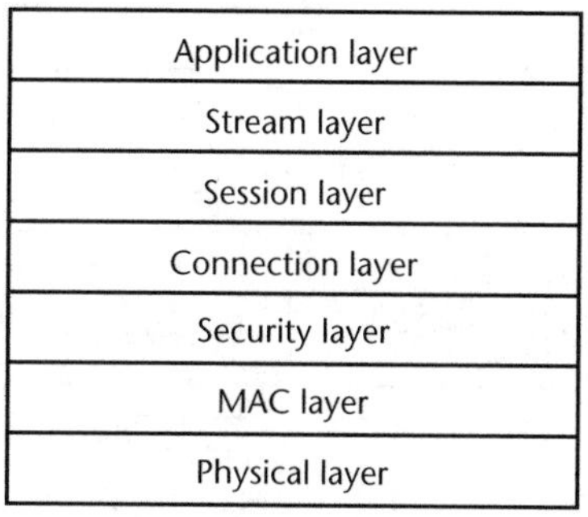

| Application layer |
| Stream layer |
| Session layer |
| Connection layer |
| Security layer |
| MAC layer |
| Physical layer |

Figure 3.14   Air interface layering architecture [11, 12]. Reproduced under written permission from Telecommunications Industry Association.

- *Stream layer:* Provides multiplexing of distinct application streams;
- *Session layer:* Provides address management, protocol negotiation, protocol configuration, and state maintenance services;
- *Connection layer:* Provides air link connection establishment and maintenance services;
- *Security layer:* Provides authentication and encryption services;
- *MAC layer:* Defines the procedures used to receive and to transmit over the physical layer;
- *Physical layer:* Provides the channel structure, frequency, power output, modulation, and encoding specifications for the forward and reverse channels.

3.3.1.3   Protocol Architecture

The protocol architecture consists of both default and nondefault protocols. Revision 0 supports only default protocols, whereas Revision A supports both default and nondefault protocols. Figures 3.15 and 3.16 present the default and nondefault protocols defined for each one of the layers, which are described in this section.

Application Layer Protocols
The application layer consists of several protocols as described below:

Default Signaling Application.   The Default signaling application encompasses the Signaling Network Protocol (SNP) and Signaling Link Protocol (SLP). The relationship between SNP and SLP is illustrated in Figure 3.17.

- The SNP provides transmission of signaling messages.
- The SLP has two layers: the delivery layer and the fragmentation layer. The purpose of the SLP delivery layer (SLP-D) is to provide reliable and best-effort message delivery for SNP packets. SLP fragmentation layer (SLP-F) provides fragmentation for SLP-D packets.

Default Packet Application.   The default packet application provides transport mechanism to carry packets between the access terminal and access network. It

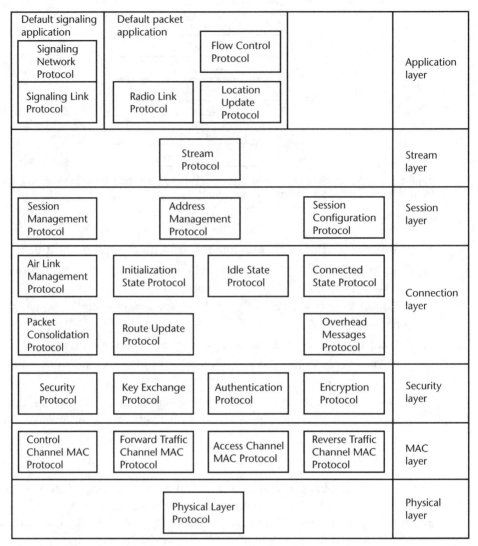

Default signaling application	Default packet application			Application layer
Signaling Network Protocol		Flow Control Protocol		
Signaling Link Protocol	Radio Link Protocol	Location Update Protocol		
	Stream Protocol			Stream layer
Session Management Protocol	Address Management Protocol		Session Configuration Protocol	Session layer
Air Link Management Protocol	Initialization State Protocol	Idle State Protocol	Connected State Protocol	Connection layer
Packet Consolidation Protocol	Route Update Protocol		Overhead Messages Protocol	
Security Protocol	Key Exchange Protocol	Authentication Protocol	Encryption Protocol	Security layer
Control Channel MAC Protocol	Forward Traffic Channel MAC Protocol	Access Channel MAC Protocol	Reverse Traffic Channel MAC Protocol	MAC layer
	Physical Layer Protocol			Physical layer

Figure 3.15 Default protocols [11, 12]. Reproduced under written permission from Telecommunications Industry Association.

consists of the Radio Link Protocol, Location Update Protocol, and Flow Control Protocol.

- *Radio Link Protocol (RLP):* Provides in-order delivery of RLP packets, retransmission, and duplicate detection, thus reducing the radio link error rate as seen by the higher layer protocols.
- *Location Update Protocol (LUP):* Defines location update procedures and messages in support of mobility management for the packet application.
- *Flow Control Protocol (FCP):* Provides flow control for the Default Packet Application Protocol.

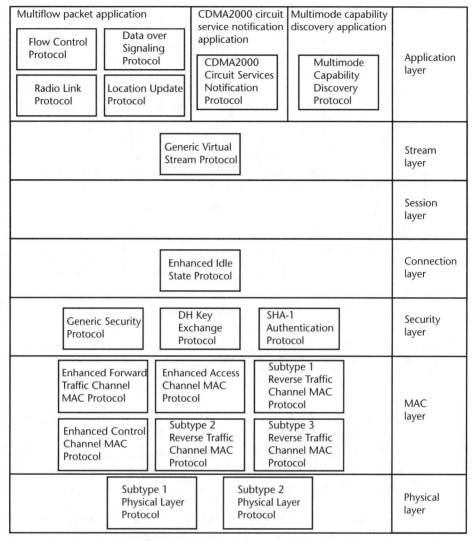

Figure 3.16 Nondefault protocols [12]. Reproduced under written permission from Telecommunications Association.

Multiflow Packet Application. The multiflow packet application (MFPA) provides multiple octet[4] streams that can be used to carry octets between the access terminal and the access network. The protocols associated with the MFPA are as follows and are illustrated in Figure 3.18 [12]:

- *RLP:* Carries one or more octet streams from the higher layer via an RLP flow. RLP uses Nak-based retransmissions.
- *Data over Signaling Protocol:* Provides transmission and duplication of higher layer packets using signaling messages. It is used to carry a small amount of data without setting up the traffic channel.

---

4. An octet is an 8-bit storage unit, often used instead of byte.

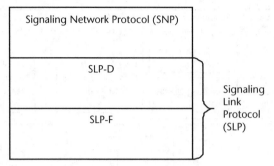

Figure 3.17   Default signaling layer protocols [11, 12]. Reproduced under written permission from Telecommunications Industry Association.

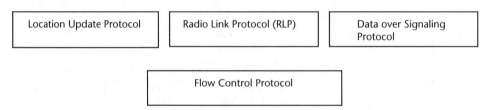

Figure 3.18   Multiflow packet application protocols [12]. Reproduced under written permission from Telecommunications Industry Association.

- *LUP:* Defines location update procedures and messages for mobility management for the MFPA.
- *FCP:* Provides procedures and messages used by the access terminal and the access network to perform flow control for the MFPA. This protocol can be in one of the following states:

  - *Close state:* In this state the MFPA does not send or receive any RLP packets or Data over Signaling Protocol messages.
  - *Open state:* In this state the MFPA can send or receive any RLP packets or Data Over Signaling Protocol messages.

CDMA2000 1X Circuit Services Notification Application.   The 3G1x circuit services notification application provides the CDMA2000 Circuit Services Notification Protocol, which sends notifications for 3G1X circuit-switched services to the access terminal through the CDMA2000 high rate packet data air interface while the access terminal is tuned to the CDMA2000 high rate packet data channel. The protocol also ensures that the access terminal stays registered in the 3G1X core circuit network even when it is monitoring the high rate packet data CDMA channel [12].

Multimode Capability Discovery Application.   The multimode capability discovery application uses the Multimode Capability Discovery Protocol to allow the AT to receive simultaneously EVDO-A and 3G1x common and traffic channels via tunable receivers. It is used in the hybrid mode (EV-DO and 3G1x) of operation [12].

Stream Layer Protocols

- *Default Stream Protocol:* This protocol provides the ability to multiplex up to four application streams. Stream 0 is always reserved for a signaling application and, by default, is assigned to the default signaling application. The stream layer uses this protocol to provide multiplexing of distinct application streams (i.e., signaling, user data) and provisioning of configuration messages that map applications to streams.
- *Generic Virtual Stream Protocol:* This protocol provides the ability to multiplex up to 127 application streams.

Session Layer Protocols

No nondefault protocol is associated with this layer, and the given protocols are applicable to both revisions.

- *Default Session Management Protocol:* Provides the means to control the activation of the Address Management and the Session Configuration Protocols. This protocol also ensures that the session is still valid and manages closing of the session.
- *Default Address Management Protocol:* Provides access terminal identifier (ATI) management and maintains the AT addresses.
- *Default Session Configuration Protocol:* Provides negotiation and configuration of the protocols and parameters used in the session.

Connection Layer Protocols

- *Default Air Link Management Protocol:* Maintains the overall connection state in the access terminal and the access network.
- *Default Overhead Messages Protocol:* The QuickConfig message and the Sector Parameters message are collectively termed as overhead messages. These messages are broadcast by the access network over the control channel. These messages are unique, in that they pertain to multiple protocols and are, therefore, specified separately. The Overhead Messages Protocol provides procedures related to transmission, reception, and supervision of these messages.
- *Default Initialization State Protocol:* Performs the actions associated with acquiring an access network.
- *Default Idle State Protocol:* Performs the actions associated with an access terminal that has acquired the network, but does not have an open connection.
- *Enhanced Idle State Protocol:* The differences between the two idle states are how access terminal transitioned from the sleep state to the monitor state and how AT requests a connection. This protocol supports shorter slot cycles for reduced page delay and longer cycles for better battery life. 3G1x and EVDO-0 systems commonly use a fixed slot cycle of 2 (5.12 seconds), while the EVDO-A slot cycle is configurable to 0.213.6, 0.427, 1.28, 2.56, and 5.12 seconds [12].

- *Default Connected State Protocol:* Performs the actions associated with an access terminal that has an open connection.
- *Default Route Update Protocol:* Protocol keeps track of an access terminal's approximate location and maintains the radio link between the access terminal and the access network.
- *Default Packet Consolidation Protocol:* Performs packet consolidation between different streams on the transmit end and packet demultiplexing on the receive end.

Security Layer Protocols

- *Default Security Protocol:* Transfers packets between the Authentication Protocol and the MAC sublayer. Security Protocol provides public variables needed by the authentication and encryption protocols (e.g., cryptosync, time stamp, and so on). This protocol does not add a header or a trailer.
- *Generic Security Protocol:* The Generic Security Protocol has a header but it does not add a trailer. It provides a cryptosync that may be used by the negotiated Authentication Protocol and Encryption Protocol on the transmission side and computes the cryptosync and makes it publicly available on the receiving side.
- *Default Key Exchange Protocol:* Provides procedures to access terminal and access network to exchange security keys for authentication and encryption.
- *DH Key Exchange Protocol:* Provides a method for session key exchange based on the Diffie-Hellman (DH) key exchange algorithm [12].
- *Default Authentication Protocol:* Provides procedures to access terminal and access network for traffic authentication.
- *SHA-1 Authentication Protocol:* The SHA-1 Authentication Protocol provides a method for authentication of the access channel MAC layer packets by applying the SHA-1 hash function to message bits that are composed of the ACAuthKey, security layer payload, CDMA system time, and the sector ID [12].
- *Default Encryption Protocol:* The Default Encryption Protocol in IS-856 does not do any encryption/decryption; that is, it does not alter the security layer packet payload. It does not add an Encryption Protocol header or trailer; therefore, the cipher text for this protocol is equal to the connection layer packet.

MAC Layer Protocols

- *Default Control Channel MAC Protocol:* Provides the procedures and messages required for an access network to transmit and for an access terminal to receive the control channel. This protocol operates with the Default (Subtype 0) Physical Layer Protocol, Subtype 1 Physical Layer Protocol, or the Subtype 2 Physical Layer Protocol [12].
- *Enhanced Control Channel MAC Protocol:* Has the same functionality as the Default Control Channel MAC Protocol, but it only operates with the Subtype 2 Physical Layer Protocol [12].

- *Default Access Channel MAC Protocol:* Provides the procedures and messages required for an access terminal to transmit and for an access network to receive the access channel. This protocol operates with the Default (Subtype 0) Physical Layer Protocol, Subtype 1 Physical Layer Protocol, or the Subtype 2 Physical Layer Protocol.

- *Enhanced Access Channel MAC Protocol:* Has the same functionality as the Default Access Channel MAC Protocol, but it does not operate with the Subtype 0 Physical Layer Protocol [12].

- *Default Forward Traffic Channel MAC Protocol:* Supports both variable rate and fixed rate of forward traffic channel. This protocol operates with the Default (Subtype 0) Physical Layer Protocol or the Subtype 1 Physical Layer Protocol.

- *Enhanced Forward Traffic Channel MAC Protocol:* Provides the procedures and messages required for an access network to transmit and for an access terminal to receive the forward traffic channel. This protocol operates with the Subtype 2 Physical Layer Protocol [12].

- *Reverse Traffic MAC Protocol:* Dictates the rules the access terminal follows to assist the access network in acquiring the reverse traffic channel. The access terminal and access network also use this protocol to determine the data rate over the reverse traffic channel. This protocol operates with the Default (Subtype 0) Physical Layer Protocol or the Subtype 1 Physical Layer Protocol.

- *Subtype 1 Reverse Traffic Channel MAC Protocol:* Provides the procedures and messages required for an access terminal to transmit, and for an access network to receive the reverse traffic channel. This protocol operates with the Default (Subtype 0) Physical Layer Protocol or the Subtype 1 Physical Layer Protocol [12].

- *Subtype 2 Reverse Traffic Channel MAC Protocol:* Has the same functionality as the Subtype 1 Reverse Traffic Channel MAC protocol. In addition, the protocol supports intra-access terminal QoS for multiple concurrent MAC flows at the access terminal by adjusting the traffic-to-pilot power ratio (T2P) control. The rate control is achieved by distributed rate selection (at the access terminal) with centralized (scheduled) resource allocation (by the access network).This protocol operates with the Default (Subtype 0) Physical Layer Protocol or the Subtype 1 Physical Layer Protocol [12].

- *Subtype 3 Reverse Traffic Channel MAC Protocol:* Has the same functionality as the Subtype 2 Reverse Traffic Channel MAC Protocol, but it operates only with Subtype 2 Physical Layer Protocol [12].

Physical Layer Protocols

- *Default (Subtype 0) and Subtype 1 Physical Layer Protocol:* Provides channel structure, frequency, power output, and modulation specifications for the forward and reverse links. A physical layer packet carries one or more MAC Layer packets. The protocol allows peak data rates of 2.4 Mbps in the forward link and 153.6 Kbps in the reverse link.

- *Subtype 2 Physical Layer Protocol:* Provides channel structure, frequency, power output, and modulation specifications for the forward and reverse links. A physical layer packet carries one MAC layer packet. The protocol allows peak data rates of 3.072 Mbps in the forward link and 1.8 Mbps in the reverse link [12].

3.3.1.4 IS-856 Operation Overview

The IS-856 system consists of a network of BSs that provide high-rate data access, on a wireless CDMA channel, to a collection of static or mobile user terminals. During an IS-856 connection, each access terminal maintains a radio link with one or more BSs, which constitute the active set of the user. If the active set of a terminal consists of more than one BS, the terminal is said to be in soft handoff. The terminal receives data from a BS in its active set on the forward link, and transmits (broadcasts) data to all the BSs in its active set on the reverse link of the IS-856 channel.

Link adaptation refers to the process of allocating/changing the transmission data rate (and possibly other resources, such as power, bandwidth, and code channels) in an efficient manner in response to channel variations. The IS-856 link adaptation consists of an open-loop rate control procedure, which determines the data rate of a packet prior to its transmission. To support link adaptation, IS-856 defines a set of physical layer packet types for data transmission on the forward link.

Each packet type offers a certain nominal data rate and nominal packet length (duration), and requires a certain minimum signal quality, for reliable reception by the user terminal. IS-856 uses incremental redundancy (hybrid ARQ) in order to refine the actual data rate of the packet while it is being transmitted. If the user terminal succeeds in decoding a packet even before it has been transmitted in its entirety, it informs the BS not to transmit the remainder of the packet.

Early termination of a packet increases its effective data rate above the nominal data rate associated with the packet type. This constitutes a closed-loop rate control procedure, which enables the terminal to compensate for the margin built into the open-loop rate requests. This margin in open-loop rate control is needed to account for unpredictable channel variations, during the transmission of a packet in the future. In addition to the link adaptation and incremental redundancy techniques that seek to optimize the throughput of a given terminal, the IS-856 network employs multiuser diversity scheduling, which seeks to optimize the overall system throughput.

In an IS-856 network, each BS provides data access to a large number of terminals using a time-division multiplex (TDM) scheme, in which the BS transmits data to, at most, one user at any given time. The BS uses an opportunistic scheduling algorithm, such as the proportional-fair scheduler, to select the recipient of the next data packet, based on the packet type requested by the different terminals in the system. The scheduler takes advantage of uncorrelated time variations of the channel seen by the different terminals in the system, and seeks to serve each user at a local peak of his channel fading process. In other words, the IS-856 forward link scheduler exploits multiuser diversity, providing enhanced system throughput (relative to a static TDM scheme), while maintaining fairness among users [15].

3.3.1.5   Physical Layer

The physical layer of 1xEV-DO Revision A is the most critical element of this technology. This air interface is different not only from 2G (IS-95) but also from IS-2000 systems. This layer provides asymmetric services; that is, the peak forward link (access network to access terminal) data rate is 3.072 Mbps, whereas the peak reverse link (access terminal to access network) is 1.8 Mbps. The physical layer provides time multiplexing of forward link channels, and the downlink transmits maximum available power all the time. The reverse link is still code division multiplexed; for instance, an IS-2000 access terminal receives full power for its time rather than sharing power across time. It also defines the physical layer channels, which are described in the next subsections.

Physical Layer Subtypes 0, 1, and 2
The IS-856 Revision A standard defines three subtypes of physical layers that are required to support a variety of data rates. Subtypes can be considered as P_Rev (Protocol Revision) for the different 1xEV-DO Revision 0/A protocols. For example, the Subtype 2 Physical Layer is necessary to support the peak data rates of 3.072 and 1.8 Mbps in the forward and reverse links, respectively. In general:

IS-856 Rev A Subtype 0 Physical Layer = IS-856 Rev 0 Physical Layer

IS-856 Rev A Subtype 1 Physical Layer = IS-856 Rev 0 Physical Layer + Support for Enhanced Access Channel MAC Protocol

IS-856 Subtype 2 Physical Layer = IS-856 Rev A Physical Layer

Default (Subtype 0) and Subtype 1 Physical Layer Packets
The transmission unit of the physical layer is a physical layer packet. A physical layer packet can be of length 256, 512, 1,024, 3,072, or 4,096 bits. The format of the physical layer packet depends on which channel it is transmitted. A physical layer packet carries one or more MAC layer packets.

Subtype 2 Physical Layer Packets
The transmission unit of the physical layer is a physical layer packet. A physical layer packet can be of length 128, 256, 512, 768, 1,024, 1,536, 2,048, 3,072, 4,096, 5,120, 6,144, 8,192, or 12,288 bits. The format of the physical layer packet depends on which channel it is transmitted. A physical layer packet carries one MAC layer packet [12].

Frame Alignment
The forward and reverse channel frames are defined in 1.66-ms slots. Each frame consists of 16 slots and is 26.67 ms in length (see Figure 3.19). Each slot contains 2,048 PN chips.

System Time
All 1xEV-DO sector air interface transmissions are referenced to a common 1xEV-DO system-wide timing reference that uses Global Positioning System (GPS) time. GPS is traceable to and synchronous with Universal Coordinated Time (UTC).

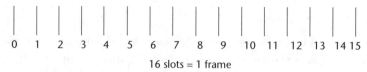

0 1 2 3 4 5 6 7 8 9 10 11 12 13 14 15

16 slots = 1 frame

Figure 3.19 1xEV-DO frame alignment.

GPS and UTC differ by an integer number of seconds, specifically the number of leap second corrections added to UTC since January 6, 1980. The start of 1xEV-DO System Time is January 6, 1980 00:00:00 UTC, which coincides with the start of GPS time.

Forward Link Channels
The 1xEV-DO forward link consists of pilot, MAC, traffic, and control channels as shown in Figure 3.20.

Pilot Channel.   The pilot channel is transmitted at all times by the sector on each active forward EV-DO channel. The pilot channel is an unmodulated signal that is used for synchronization and other functions by an access terminal operating within the coverage area of the sector. The pilot channel is transmitted at the full sector power.

MAC Channel.   The MAC channel is composed of Walsh channels that are orthogonally covered and binary phase shift keying (BPSK) modulated on a particular phase of the carrier (either in-phase or quadrature phase). Each Walsh channel is identified by a MACIndex value that is between 0 and 63 and defines a unique 64-ary Walsh cover and a unique modulation phase. In the subtype 2 physical layer, each Walsh channel is identified by a MACIndex value that is between 0 and 127 and defines a unique 128-ary Walsh cover and a unique modulation phase.

- *Reverse power control (RPC) channel:* Used for transmission of the RPC bit stream for a particular access terminal.
- *DRC lock channel:* The data rate control (DRC) lock channel is used for the transmission of the DRC lock bit stream destined to the access terminal [12].
- *Reverse activity (RA) channel:* The reverse activity bit transmitted by the RA channel gives an indication to the access terminal about whether to increase or decrease the rate on the reverse traffic channel.

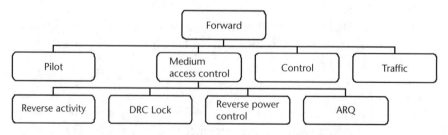

Figure 3.20 Forward link channels [12]. Reproduced under written permission from Telecommunications Industry Association.

- *Automatic repeat request (ARQ) channel:* Each sector of an access network transmits a positive acknowledgment (ACK) or a negative acknowledgment (NAK) in response to a physical layer packet using the ARQ channel. The H-ARQ, L-ARQ, and P-ARQ bits are transmitted on the ARQ channel (for subtype 2 only). The hybrid ARQ (H-ARQ) bit is sent on the ARQ channel in response to the first, second, and third subpackets of the reverse link physical packet to support the physical layer ARQ. The last ARQ (L-ARQ) bit is sent on the ARQ channel in response to the fourth (last) subpacket of the reverse link physical packet to support the MAC layer ARQ. The packet ARQ (P-ARQ) bit is sent on the ARQ channel in response to a reverse link physical packet to support the MAC layer ARQ [12].

Traffic Channel. The forward traffic channel is a packet-based, variable-rate channel. The specific data rates in the forward link are 38.4, 76.8, 153.6, 307.2, 614.4, 921.6, 1,228.8, 1,843.2, and 2,457.6 Kbps for subtypes 0 and 1. For subtype 2 the date rate varies from 4.8 Kbps to 3.072 Mbps.

Control Channel. The control channel transmits broadcast messages and access-terminal-directed messages. The control channel messages are transmitted at a data rate of 38.4 or 76.8 Kbps.

Forward Link Modulated Waveform

Downlink Modulation for Subtypes 0 and 1: Similar to IS-95 and IS-2000 systems, the 1xEV-DO forward link carriers are allocated 1.25 MHz of bandwidth and use a DS spread waveform at the rate of 1.2288 Mcps. As shown in Figure 3.21, the fundamental timing unit for forward link transmissions is a 1.666... ms slot (2048 chips per slot) that contains the Pilot and MAC channels and a data portion that may contain a traffic or control channel [11, 12, 15, 16].

The pilot channel is transmitted with 96 chips every half-slot 9 (Figure 3.22), providing not only a reference for coherent demodulation of traffic and MAC channels but also a 1,200-Hz sampling of the wireless channel state. These samples are used to estimate and predict the received signal-to-noise ratio (SNR) at the AT that is used to determine the maximum data rate that the AT can receive on the forward link. This provides the system with a mechanism for fast adaptation of modulation and coding schemes to different mobile channel environments. The RPC channel and DRC lock channel are time-division multiplexed and transmitted on the same MAC channel. Each of these channels carries 1 bit of information per slot and is BPSK modulated using a 64-ary orthogonal function. The orthogonal sequence is then repeated four times, resulting in 256 allocated chips per slot.

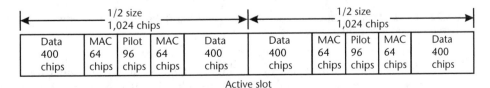

Data 400 chips	MAC 64 chips	Pilot 96 chips	MAC 64 chips	Data 400 chips	Data 400 chips	MAC 64 chips	Pilot 96 chips	MAC 64 chips	Data 400 chips

1/2 size 1,024 chips — 1/2 size 1,024 chips

Active slot

Figure 3.21 Forward link active slot structure for subtypes 0, 1, and 2 [12]. Reproduced under written permission from Telecommunications Industry Association.

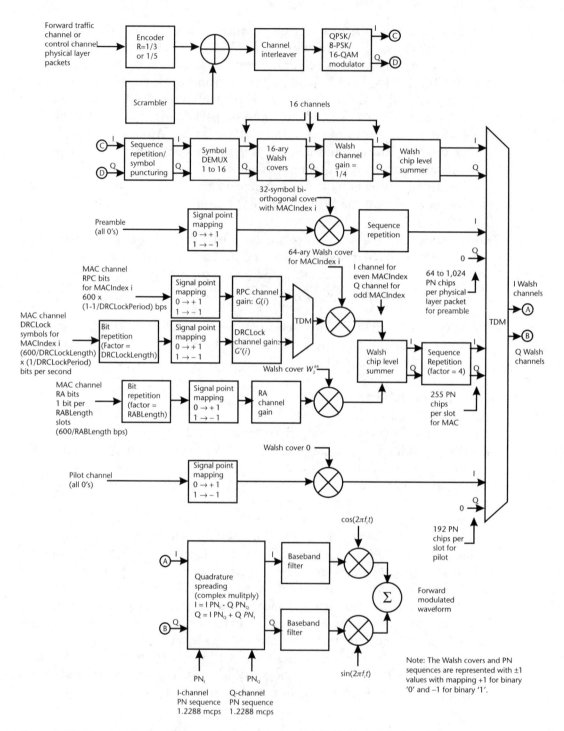

Figure 3.22 Forward channel structure for subtypes 0 and 1 [12]. Reproduced under written permission from Telecommunications Industry Association.

The traffic channel is transmitted to a single user at a time following a data rate control (DRC) message sent on the reverse link. The DRC indicates not only the data rate, but also the modulation, coding rate, preamble length, and maximum number of slots needed to transmit a physical layer packet to an AT. The modulation parameters for each of the possible DRC rates are described in Table 3.2.

The forward traffic channel and control channel data are encoded in blocks called physical layer packets. The physical layer packets are encoded with code rates of $R = 1/3$ or $1/5$ using a parallel turbo encoder. The output of the encoder is scrambled and then fed into a channel interleaver. The output of the channel interleaver is fed into a QPSK/8-PSK/16-QAM modulator. The modulated symbol sequences are repeated and punctured, as necessary. Then, the resulting sequences of modulation symbols are demultiplexed to form 16 pairs (in-phase and quadrature) of parallel streams.

Each of the parallel streams is covered with a distinct 16-ary Walsh function at a chiprate to yield Walsh symbols at 76.8 kilosymbols per second (Ksps). The Walsh-coded symbols of all the streams are summed to form a single in-phase stream and a single quadrature stream at a chiprate of 1.2288 Mcps. The resulting chips are time-division multiplexed with the preamble, pilot channel, and MAC channel chips to form the resultant sequence of chips for the quadrature spreading operation. Following the quadrature spreading, the I′ and Q′ impulses are applied to the inputs of the I and Q baseband filters. After filtering, the two sequences are combined to form the forward modulated waveform.

Another important aspect of physical layer packet transmissions is the multislot interlacing approach, which is necessary to support the adaptive rate control on the forward link. This is exemplified in Figure 3.23, where a 153.6-Kbps packet is requested (DRC) and the corresponding four slots are transmitted with three slots of separation between them. This allows time for the receiving AT to attempt decoding of the partially received packet and to indicate to the transmitting sector (via the ACK channel on the reverse link) whether or not the decoding attempt was successful. Note that there is a fixed relation between the timing of a received DRC or ACK on the reverse link and their corresponding effects on the forward link

Table 3.2 Modulation Parameters of the Forward Link for Subtypes 0 and 1 [11]

Data Rate (Kbps)	Number of Slots per Packet	Information Bits	Turbo Code Rates	Modulation	Preamble Length in Chips
38.4	16	1,024	1/5	QPSK	1,024
76.8	8	1,024	1/5	QPSK	512
153.6	4	1,024	1/5	QPSK	256
307.2	2	1,024	1/5	QPSK	128
307.2	4	2,048	1/3	QPSK	64
614.4	1	1,024	1/3	QPSK	64
614.4	2	2,048	1/3	QPSK	64
921.6	2	3,072	1/3	8PSK	64
1,228.8	1	2,048	1/3	QPSK	64
1,228.8	2	4,096	1/3	16QAM	64
1,843.2	1	3,072	1/3	8PSK	64
2,457.6	1	4,096	1/3	16QAM	64

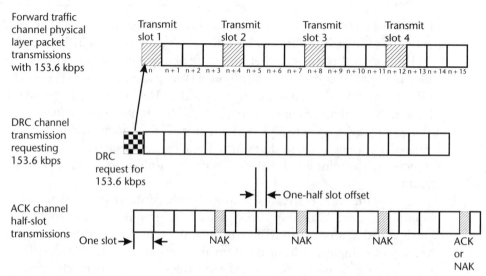

Figure 3.23 Multislot physical packet with normal termination for subtypes 0, 1, and 2 [11, 12].

transmissions. In addition, simultaneous packet transmissions to single or different ATs may be interlaced in time.

Downlink Modulation for Subtypes 2.   As stated before, the IS-856 subtype 2 physical layer is the new IS-856 Revision A physical layer that is required to support the peak data rates. The overall subtype 2 modulation process (Figure 3.24) is very similar to the modulated waveform of subtypes 0 and 1. Although the process is similar there are some differences, and those are discussed next.

The MAC channel consists of an additional ARQ channel. Each MAC channel is QPSK modulated on one of 128 128-ary Walsh codewords (covers). The MAC symbol Walsh covers should be transmitted two times per slot in the four bursts of 64 chips each. This results in 256 PN chips per slot for MAC. The RPC channel and the DRC lock channel are transmitted on the in-phase and quadrature phase of the same MAC channel. The RPC channel and the DRC lock channel are time-division multiplexed with the ARQ channel and transmitted in one out of every four slots.

Forward traffic channel and control channel physical layer packets are transmitted in 1 to 16 slots as shown in Table 3.3 (partial table is shown). When more than one slot is allocated, the transmit slots use four-slot interlacing.

Similar to that for the subtypes 0 and 1 physical layer, the resulting chips of the forward traffic channel and control channel are time-division multiplexed with the preamble, pilot channel, and MAC channel chips to form the resultant sequence of chips for the quadrature spreading operation. Following the quadrature spreading, the I′ and Q′ impulses are applied to the inputs of the I and Q baseband filters. After filtering, the two sequences are combined to form the forward modulated waveform.

Reverse Link Channels
The reverse link of IS-856 is quite similar to the reverse link in an IS-2000 system. The IS-856 reverse channel structure, as described in Figure 3.25, consists of two main channels, access and traffic.

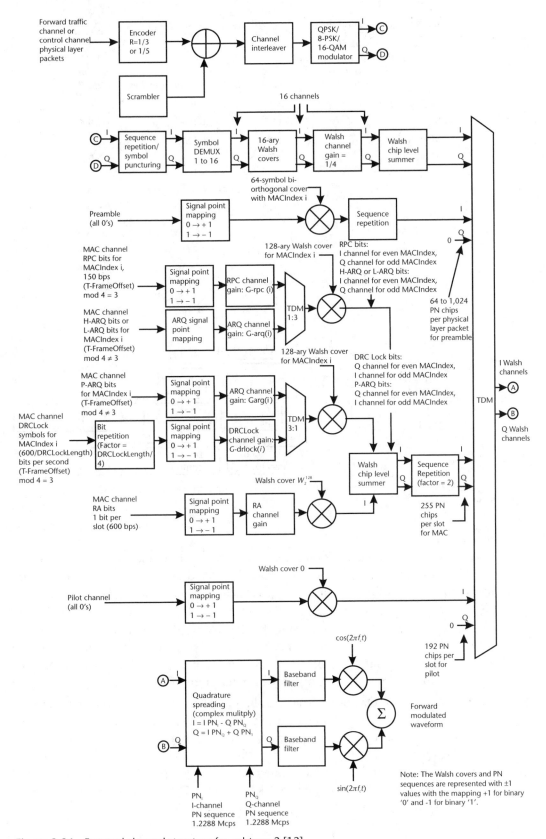

Figure 3.24   Forward channel structure for subtype 2 [12].

Table 3.3   Modulation Parameters of the Forward Link for Subtype 2 [12]

Transmission Format Physical Layer Packet Size (bits), Nominal Transmission Duration (slots), Preamble Length (chips)	Code Rate	Modulation Type	Nominal Data Rate
(128, 16, 1024)	1/5	QPSK	4.8
(128, 8, 512)	1/5	QPSK	9.6
(128, 4, 1024)	1/5	QPSK	19.2
(128, 4, 256)	1/5	QPSK	19.2
:			
:			
(4096, 2, 64)	1/3	16-QAM	1,228.8
(4096, 1, 64)	1/3	16-QAM	2,457.6
(5120, 2, 64)	1/3	16-QAM	1,536.0
(5120, 1, 64)	1/3	16-QAM	3,072.0

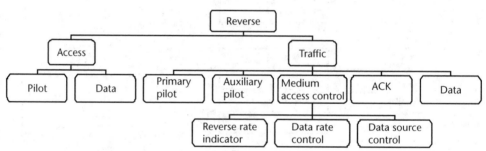

Figure 3.25   Reverse link channels [9]. Reproduced with permission of The McGraw-Hill Companies.

Access Channel.   The access terminal to initiate communication with the access network or to respond to an access terminal directed message uses the access channel. It consists of a pilot channel and a data channel as shown in Figure 3.25.

- *Pilot channel:* The access terminal will transmit unmodulated symbols with a binary value of "0" on the pilot channel. The pilot channel will be transmitted continuously during access terminal transmission. It will be transmitted on the 'I' channel using the 16-chip Walsh function number 0 cover.

- *Data channel:* One or more access channel physical layer packets will be transmitted on the data channel during every access probe. The access channel physical layer packets can be transmitted at a fixed data rate of 9.6 Kbps on the 'Q' channel using the four-chip Walsh function number 2. For the subtype 2 physical layer, the access channel physical layer packets can be transmitted at data rates of 9.6, 19.2, or 38.4 Kbps on the Q' channel using the four-chip Walsh function number 2.

Reverse Traffic Channel.   The access terminal to transmit user-specific traffic or signaling information to the access network uses the reverse traffic channel. It consists of a pilot channel, a reverse rate indicator (RRI) channel, a DRC channel, an ACK channel, and a data channel. As part of subtype 2, the reverse traffic channel also includes a data source control channel and an auxiliary pilot channel [9].

- *Pilot channel:* The access terminal will transmit unmodulated symbols with a binary value of "0" on the pilot channel.
- *MAC channel:* The MAC channel consists of RRI and DRC channels. As part of subtype 2, it also includes the data source control (DSC) channel [9].
- *RRI channel:* The access terminal uses the RRI channel to indicate the data rate transmitted on the reverse traffic channel. As part of subtype 2, the RRI channel is used by the access terminal to indicate the payload size and subpacket identifier of the physical layer packet transmitted on the data channel [12].
- *DRC channel:* The DRC channel is used by the access terminal to indicate to the access network the selected serving sector and the requested data rate on the forward link.
- *DSC channel:* The access terminal uses the DSC channel to indicate to the access network the selected serving cell on the forward link. The serving cell is indicated by the DSC value (3-bit value) for that cell [12].
- *ACK channel:* The ACK channel is used by the access terminal to inform the access network whether a data packet transmitted on the forward traffic channel has been received successfully or not.
- *Data channel:* The data channel is transmitted at data rates of 9.6, 19.2, 38.6, 76.8, or 153.6 Kbps for subtypes 0 and 1. All data transmitted on the reverse traffic channel is encoded, block interleaved, sequence repeated, BPSK modulated, and orthogonally spread by Walsh function W42 for subtypes 0 and 1. For subtype 2 the data channel is transmitted at data rates ranging from 4.8 Kbps to 1.8432 Mbps. All data transmitted on the reverse traffic channel is encoded, block interleaved, sequence repeated, BPSK, QPSK, or 8-PSK modulated, and orthogonally spread by Walsh function W24 or Walsh function W12 or both. The data channel is modulated as shown in Table 3.4.
- *Auxiliary pilot channel:* The auxiliary pilot channel is transmitted during the nth half-slot if the transmitted payload during half-slots $n - 1$ or $n + 1$ is greater than or equal to AuxiliaryPilotChannelMinPayload, which is public data of the Reverse Traffic Channel MAC Protocol. If the auxiliary pilot

Table 3.4   Data Channels Modulation Format [12]

Modulation Format	(Modulation, Walsh Function)
B4	$(BPSK, W_2^4)$
Q4	$(QPSK, W_2^4)$
Q2	$(QPSK, W_1^2)$
Q4Q2	$(QPSK, W_2^4) + (QPSK, W_1^2)$
E4E2	$(8\text{-PSK}, W_2^4) + (8\text{-PSK}, W_1^2)$

channel is transmitted during the nth half-slot, its power shall be specified relative to the maximum of the data channel gains during half slots $n - 1$ and $n + 1$ by AuxPilotChannelGain, where AuxPilotChannelGain is public data of the reverse traffic channel MAC Protocol. The auxiliary pilot channel power level shall be updated at the start of every half-slot [12].

Reverse Link Modulated Waveform

Uplink Modulation for Subtypes 0 and 1.   Figures 3.26 and 3.27 show the reverse traffic channel block diagram of the IS-856 standard. There are four orthogonal code-division multiplexed channels. The pilot/RRI channel is time multiplexed so that the RRI channel is transmitted during 256 chips at the beginning of every slot (1.66… ms). The 3-bit RRI symbol transmitted every frame (16 slots) is encoded using a 7-bit simplex codeword [11, 17].

Each codeword is repeated 37 times over the duration of the frame, while the last 3 code symbols are not transmitted. The DRC symbols (4 bits indicating the

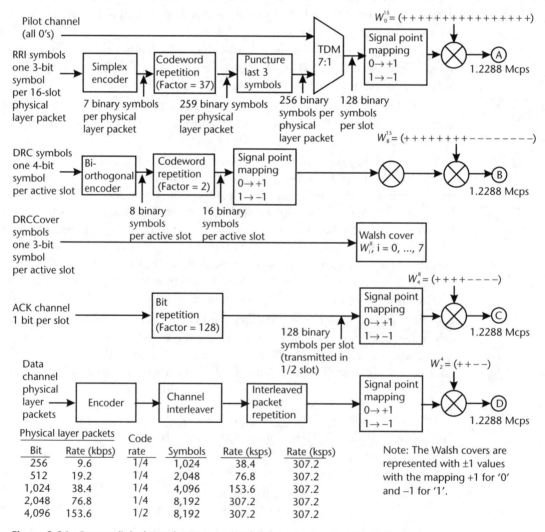

Figure 3.26 Reverse link channels structure (1 of 2) for subtypes 0 and 1 [12].

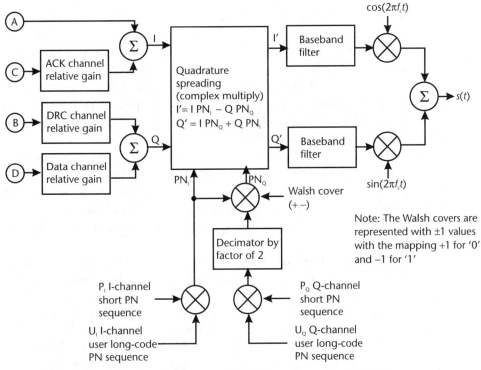

Figure 3.27 Reverse link channels structure (2 of 2) for subtypes 0 and 1 [12].

desired data rate) are encoded using 16-ary biorthogonal code. Each code symbol is further spread by one of the 8-ary Walsh functions indicating the desired transmitting sector on the forward link. The DRC message is transmitted half-slot offset with respect to a slot boundary. The reason is to minimize prediction delay while providing enough time for processing at the desired sector before transmission on the forward link starts on the next slot. The ACK channel is BPSK modulated in the first half-slot (1,024 chips) on an active slot. The transmissions on the ACK channel only occur if the access terminal detects a data packet directed to it on the forward traffic channel. For a forward data packet transmitted in slot n, a 0 bit is transmitted on the ACK channel in slot $n + 3$ if a data packet transmission has been successful; otherwise, a 1 bit is transmitted. The three slots of delay allow the terminal to demodulate and decode the received packet before transmitting on the ACK channel.

The data channel supports data rates of 9.6, 19.2, 38.4, 76.8, and 153.6 Kbps with 16-slot packets (26.66... ms) with BPSK modulation. The packet is encoded (turbo encoder) using either 1/2 or 1/4 rate parallel code. The code symbols are bit-reversal interleaved and block repeated, achieving a fixed 307.2-Ksps modulation symbol rate. The pilot/RRI, DRC, ACK, and data channel modulation symbols are each spread by an appropriate orthogonal Walsh function. Before quadrature spreading, the pilot/RRI and ACK channels are scaled and combined to form the in-phase component. Similarly, the data and DRC channels are scaled and combined to form the quadrature component of the baseband signal. Similar to the forward link after quadrature spreading, the I′ and Q′ impulses are applied to the inputs of the I and Q baseband filters. After filtering, the two sequences are combined to form the reverse modulated waveform.

Reverse link power control (both open and closed loops) is applied to the pilot/RRI channel only. The power allocated to the DRC, ACK, and data channels are adjusted by a fixed gain relative to the pilot/RRI channel in order to guarantee the desired performance of these channels. For example, the relative gain of the data channel increases with the data rate so that the received E_b/N_t is adjusted to achieve the required packet error rate (PER).

Uplink Modulation for Subtype 2 Physical Layer.   Figures 3.28 and 3.29 show the reverse traffic channel block diagram of the IS-856-A subtype 2 physical layer. Besides the subtype 0 and 1 channels, subtype 2 also includes an auxiliary pilot channel and a DSC. All channels are spread with Walsh functions, called Walsh covers, at a fixed rate of 1.228 Mcps [12].

The 6-bit RRI symbol per four-slot physical layer subpacket is formed using the 4-bit symbol representing the payload size and the 2-bit symbol representing the subpacket identifier. The 6-bit RRI symbol is encoded using a 32-dimensional biorthogonal signal constellation, repeated with a factor of 4 and then spread with Walsh function 4. The ACK channel is BPSK modulated in the first half-slot (1,024 chips) and is transmitted on the 'I' channel. The DRC symbols (4 bits indicating the desired data rate) are encoded using 8-bit orthogonal code. Each code symbol is further spread by one of the 8-ary Walsh functions indicating the desired transmitting sector on the forward link. The 3-bit DSC symbols (indicating the serving cell) take effect one slot after the end of its transmission and stay in effect for DSCLength slots. The DSC values are transmitted at a data rate of 600/DSCLength values per second. These are block encoded to form 32-bit codewords, repeated, spread by a 32-ary Walsh function, and transmitted on the in-phase (I) channel.

The data channel supports data rates ranging from 4.8 Kbps to 1.8432 Mbps. The packet is encoded (turbo encoder) using different codes as shown in Table 3.5 (partial table is shown). The code symbols are scrambled, interleaved, and modulated. The data channel is BPSK, QPSK, or 8-PSK modulated and each modulated stream is covered by either a 4-ary Walsh function or a 2-ary Walsh function. The data channel is B4 modulated (BPSK modulation with 4-ary Walsh cover), Q4 modulated (QPSK modulation with 4-ary Walsh cover), Q2 modulated (QPSK modulation with 2-ary Walsh cover), Q4Q2 modulated (sum of the Q4 and Q2 modulated symbols), or E4E2 modulated (sum of E4 (8-PSK modulated with 4-ary Walsh cover), and E2 (8-PSK modulated with 2-ary Walsh cover) modulated symbols. Before quadrature spreading, the pilot, auxiliary pilot, ACK, and DSC channels and in-phase of the data channel are scaled and combined to form the in-phase component. The DSC and ACK are also time multiplexed before combining with the rest of the channels to form the in-phase.

Similarly, the quadrature phase of the data channel and the DRC channel are scaled and combined to form the quadrature component of the baseband signal. Similar to the forward link after quadrature spreading, the I′ and Q′ impulses are applied to the inputs of the I and Q baseband filters. After filtering, the two sequences are combined to form the reverse modulated waveform.

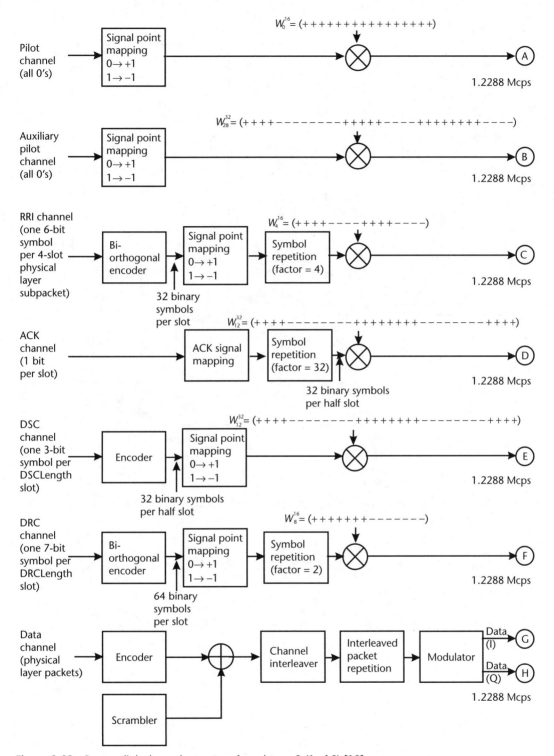

Figure 3.28 Reverse link channels structure for subtype 2 (1 of 2) [12].

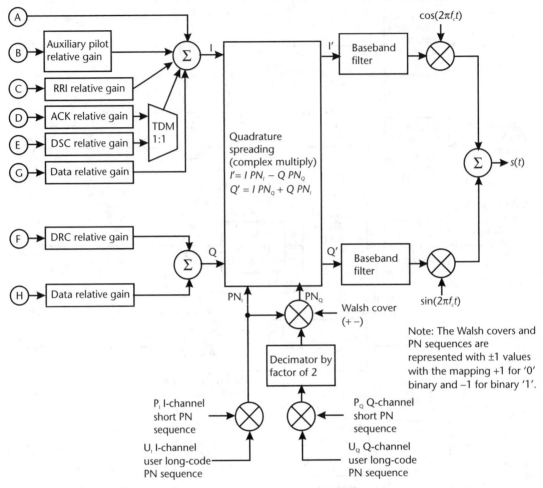

Figure 3.29   Reverse link channels structure for subtype 2 (2 of 2) [12].

Table 3.5   Modulation Parameters for the Reverse Traffic Channel [12]

Payload Size (bits)	Modulation	Effective Data Rate (Kbps)				Code Rate (Repetition)			
		After 4 Slots	After 8 Slots	After 12 Slots	After 16 Slots	After 4 Slots	After 8 Slots	After 12 Slots	After 16 Slots
128	B4	19.2	9.6	6.4	4.8	1/5 [3.2]	1/5 [6.4]	1/5 [9.6]	1/5 [12.8]
256	B4	38.4	19.2	12.8	9.6	1/5 [1.6]	1/5 [3.2]	1/5 [4.8]	1/5 [6.4]
		⋮							
		⋮							
8,192	Q4Q2	1,228.8	614.4	409.6	409.6	2/3 [1]	1/3 [1]	2/9 [1]	1/5 [1.2]
12,288	E4E2	1,843.2	921.6	614.4	460.8	2/3 [1]	1/3 [1]	1/3 [1.5]	1/3 [2]

3.3.2   CDMA2000 1xEV-DV

CDMA2000 1xEV-DV is based on the IS-2000 Release D standard and was published in March 2004. Unlike 1xEV-DO it supports both voice and data and is backward compatible with 1xRTT. There is only one mandatory feature in the standard Mobile Equipment Identification (MEID). MEID is a 52-bit mobile identification number that replaces the existing 32-bit Electronic Serial Number (ESN). This mandatory feature is required to avoid the mobile ID exhaustion that currently threatens the entire wireless community [13, 14].

3.3.2.1 1xEV-DV Architecture

Figure 3.30 depicts the CDMA2000 Release D Open System Interconnection (OSI) architecture, which is very similar to CDMA2000 1X. The enhancements occur at the physical layer and are controlled by the upper layers. This section describes the 1xEV-DV air interface architecture by examining the new channels.

1xEV-DV Channels and Channel Configurations
1xEV-DV added few new channels and radio configurations to the existing CDMA2000 1X architecture to support the enhancements.

- *Packet data channel (PDCH):* PDCH is the main packet data channel for supporting high data rates using H-ARQ, adaptive modulation and coding (AMC), and time- and code-division multiplexing schemes. Each BS can transmit up to two PDCHs or one per traffic channel using radio configuration 10 (a new RC).

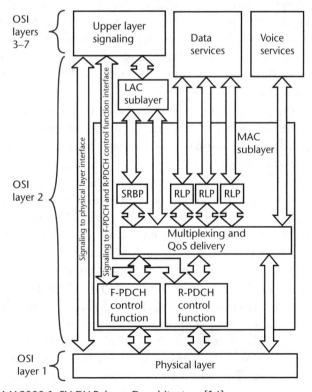

Figure 3.30 CDMA2000 1xEV-DV Release D architecture [14].

- *Forward packet data control channel (F-PDCCH):* F-PDCCH is used by the BS for transmitting control information for the associated F-PDCCH or transmitting a Walsh mask to the MS with SR1. A forward CDMA channel may contain up to two F-PDCCHs. It typically sends demodulation, decoding, and ARQ information to the specific mobiles.
- *Forward grant channel:* The BS uses this channel to grant the MS operating with RS1 permission to transmit on R-PDCCH.
- *Forward indicator control channel:* Consists of indicator control subchannels. When the subchannel is used for rate control, it is called a rate control subchannel and when used for power it is called a common power control subchannnel.
- *Forward ACK channel:* This is an ON-OFF channel that provides ACK or NAK responses for the R-PDCH transmissions.
- *Reverse channel quality indicator channel (R-CQICH):* The MS informs the BS about the channel quality measurements of the best serving sector that uses R-CQICH. The mobile indicates selection of the serving sector by means of a Walsh cover. This feedback is used as an input to forward link modulation, coding, and scheduling processes.
- *Reverse ACK channel (R-ACKCH):* The MS informs the BS whether or not the F-PDCCH packet was successfully decoded by using the reverse acknowledgment channel. R-ACKCH is only used in the F-PDCCH mode. An open reverse channel is also maintained for R-ACKCH AND R-CQICH while using the F-PDCCH.
- *Reverse secondary pilot channel (R-SPCH):* The R-SPCH is an unmodulated spread-spectrum signal used to assist the BS in detecting a MS transmission on the reverse packet data channel. The secondary pilot is not transmitted unless instructed by the MAC layer. Typically the reverse secondary pilot channel is transmitted only in conjunction with the higher encoder packet sizes on the reverse packet data channel.
- *Reverse packet data control channel (R-PDCCH):* The R-PDCCH is used by the MS for transmitting control information for the associated reverse packet data channel. It consists of 6 bits and a MS indicator bit that is used at the output of the encoder bit.
- *Reverse request channel (R-RCH):* The R-RCH is used by the MS to update the BS about the amount of data in the buffer, the QoS requirements, and the available headroom. The reverse request channel is also used by the MS to request permission from the BS to transmit above the autonomous data rate on the reverse packet data channel. A R-RCH frame consists of 12 information bits, followed by a 12-bit frame quality indicator and 8 encoder tail bits.
- *Reverse packet data channel (R-PDCH):* R-PDCH is used for the transmission of user information to the BSs by the MSs operating with SR1. It will use radio configuration 7 to support encoder packet sizes of 192, 408, 792, 1,560, 3,096, 4,632, 6,168, 9,240, 12,312, and 15,384 bits. R-PDCH will support data rates from 19.2 Kbps to 1.5384 Mbps using either BPSK or QPSK modulation types. The modulation scheme will be selected based on the encoder packet sizes.

- *Hybrid automatic repeat request (H-ARQ):* The H-ARQ scheme uses a code designed for both error detection and correction, and thus it requires more parity bits (overhead) than a code used for error detection only. H-ARQ improves throughput by combining rather than discarding failed transmission attempts with the current attempt. A fast retransmission of frames received in error is critical to maintaining high bandwidth, so ARQ has migrated from the MAC layer to the physical layer.
- *Channel configurations:* Release D added several new channel configurations to support voice and data services (Table 3.6).

3.4 CDMA2000 Key Attributes

In this section we summarize some of the key features of the IS-2000 and IS-856 standards.

3.4.1 IS-2000 Enhancements

1xRTT offers a highly efficient air interface with the following key characteristics:

- *Circuit-switched voice capacity gains:* IS-2000 Release 0 provides new techniques to improve circuit-switched technology but it does not support VoIP. These systems have increased the voice capacity (i.e., the number of simultaneous voice calls in the downlink) by approximately 1.2 to 2.0 times compared to that of IS-95 systems. This increase was made possible with the introduction of techniques such as 800-Hz fast forward power control; QPSK modulation in the forward link, which doubles the number of Walsh codes; and a reverse dedicated pilot, which provides a coherent phase reference for the BS during demodulation of the reverse link signal.
- *VoIP:* IS-2000 Releases C/D are the only two releases that support VoIP. VoIP is packetized voice unlike traditional circuit-switched voice. Studies have shown that it provides roughly the same quality of service and capacity as the native voice. Currently, CDMA2000 operators are evaluating VoIP technology and have not deployed it in their networks.

Table 3.6 1xEV-DV Channel Configurations

Channel Number	1xEV-DV Channel Configurations	Basic Use
Channel configuration 0	F-PDCH + R-FPDCH	Data-only service
Channel configuration 1	F-PDCH + + R-PDCH + R-FCH	Data-only service
Channel configuration 2	F-PDCH + R-PDCH + R-DCCH	Data-only service
Channel configuration 3	F-PDCH + R-PDCH + F/R-FCH	Mixed voice and data services
Channel configuration 4	F-PDCH + R-PDCH + F-DCCH + R-DCCH	Mixed voice and data services
Channel configuration 5	F-PDCH + R-PDCH + F/R FCH + R DCCH	Mixed voice and data services
Channel configuration 6	F-PDCH + R-PDCH + F/R FCH + F/R DCCH	Mixed voice and data services

- *Simultaneous voice and data:* Except for IS-2000 Release 0, all other releases support concurrent voice and data sessions. The 1xEV-DV version utilizes the fundamental channel for traditional voice, whereas the packet data channel is used for data services.

- *Data capacity and data rates:* IS-2000 introduced the enhanced LAC and MAC layers within the link layer of the air interface to support high data rates and data throughputs. In addition, turbo codes were introduced that generated better data throughputs than convolutional codes. In particular, 1xEV-DV uses the following features to support higher data rates up to 3.09 Mbps in the downlink and 1.8 Mbps in the uplink:

  - *Adaptive modulation:* The link modulation and coding are varied in real time to adapt to the changing RF conditions. In the link adaptation scheme, the BS assigns users the best modulation and coding rate for instantaneous channel conditions.

  - *TDM and CDM multiplexing:* 1xEV-DV uses TDM and CDM to transmit different types of data. TDM is favored because it works best with FTP; however, CDM is preferred where full-frame efficiency is required for VoIP and audio/video streaming.

  - *Hybrid ARQ:* H-ARQ improves throughput by combining rather than discarding failed transmission attempts with the current attempt. A fast retransmission of frames received in error is critical to maintaining high bandwidth, so ARQ has migrated from the MAC layer to the physical layer.

- *Fast data call setup:* Each of the revisions of IS-2000 since Release 0 has made provisions for improving the call setup times for packet data sessions. With Release D of the standard, the RF call setup time is reduced from the current time of 3 to 5 seconds to less than 1 second.

- *Broadcast/multicast services (BCMCS):* Broadcast and multicast services are another feature of Release D that provides new capabilities for the operator to offer services on shared broadcast or multicast channels, providing the operator with efficient support for services such as music and streaming video without allocating dedicated resources to multiple users.

- *QoS:* QoS has been loosely defined in the standards, however, in Release D, QoS takes on new meaning as service differentiation, and packet priority can be folded into the mix to support different QoS parameters by application and user.

- *Extended battery life:* The ITU requirement for extended battery life is met with the introduction of the forward quick paging channel (F-QPCH). The goal of the F-QPCH is to greatly reduce the amount of time spent monitoring the slots of the paging channel (i.e., standby time). The channel contains a single-bit message, a Quick Page message, to direct a slotted-mode MS to monitor the assigned slot on the common paging channel that immediately follows. The Quick Page message is sent up to 80 ms before the Page message to alert the MS to listen to the paging channel. The quick paging channel uses a different modulation, so it will appear as a different physical channel. No error correction coding or interleaving is performed on this channel.

- *Backward compatibility:* Backward compatibility is defined as the ability to support the new technology, while still being able to fully support the legacy technology. In this case, backward compatibility allows service providers to support 3G1x mobiles, while still being able to support the legacy 2G mobiles, within the same cell/sector. To provide backward compatibility, the network must support both the 3G1x and 2G technologies, so that mobiles of both generations are supported in a particular cell. Conversely, new 3G1x mobiles must support both generations, to ensure operability when traveling into markets that have not yet upgraded to 3G1x capability. To be backward compatible with existing CDMAOne systems, the CDMA2000 radio interface retains many of the attributes of the CDMAOne air interface design such as a chip rate of 1.2288 Mcps, frame structure, and forward pilot channel.

3.4.2 IS-856 Enhancements

IS-856 Revision A and IS-2000 Release D have very similar technical attributes, except the former does not support circuit-switched voice. Some of the key enhancements of the two revisions of the IS-856 are as follows:

- *High-speed forward and reverse data:* 1xEV-DO Revision A provides enhancements in both software and hardware to increase the data speeds on the downlink from the current 2.45 Mbps (Revision 0) to 3.09 Mbps, and in the uplink from 153.6 Kbps (Revision 0) to 1.8 Mbps. 1xEV-DO-A incorporates most of the enhancements introduced by 1xEV-DV Release D to match the data rates of 1xEVDV.

- *Multiuser packets:* 1xEV-DO Revision A includes the ability to multiplex multiple small user packets (such as VoIP frames). A single 1,024-byte (or larger) physical layer packet can be multiplexed, leading to higher granularity and better use of the physical layer.

- *Multiflow RLP:* Multiflow RLP allows the 1xEV-DO system to send multiple RLP flows to the same subscriber and apply different QoS treatment to the flows (e.g., one flow for Internet browsing with no QoS, the second flow for VoIP with high QoS requirements).

- *QoS:* QoS is enabled in 1xEV-DO with the multiflow RLP. With this type of solution, different QoS parameters can be applied to data flow for several users.

- *Improved operation with IS-2000:* 1xEV-DO Revision A introduces new overhead signaling and messaging to support multimode devices (IS-2000/IS-856 dual-mode operation).

- *Battery life:* 1xEV-DO Revision A introduces a slotted-mode operation for the 1xEV-DO terminal, which improves its battery life.

• *Backward compatibility:* Revision A systems are backward compatible with Revision 0 systems.

3.5 3GPP2-Based Network Architecture

As stated earlier, CDMA2000 corresponds to the family of 3G standards defined by the 3GPP2. CDMA2000 includes both IS-2000- and IS-856-based networks, and multiple releases are associated with both standards. The very first CDMA2000 system was based on IS-2000 Release 0 and was called CDMA2000 1X or 1xRTT or 3G1x. Five releases of IS-2000 have been published so far: Release 0 and Releases A/B/C/D. The most significant one was Release 0, whereas releases A, B, and C did not get much industry support. Also, although Release D provided a variety of enhancements, it was outclassed by IS-856 Revision A, which provided similar capabilities. The network migration from Release 0 to Release D was easy because the former is backward compatible, which makes the transition quite simple. IS-856 has two revisions: Revision 0 and Revision A. Similarly, 1xEV-DO Revision 0 is backward compatible with Revision A. This backward compatibility minimizes the introduction of new hardware elements and most of the job is completed via newer software loads. But IS-2000 and IS-856 systems are not backward compatible with each other.

3.5.1 IS-2000 Architecture

Because all of the IS-2000 releases are backward compatible with each other, the migration from one release to the other is quite simple compared to 3GPP. The migration from IS-95 to IS-2000 involved changes to the radio access network (RAN), a new packet data serving node (PDSN), and a new authentication, authorization, and accounting (AAA) server. Once the network is upgraded from CDMAOne to CDMA2000, the migration from CDMA2000 1X to 1xEV-DV mainly requires a new 1xEV-DV channel element in the base transceiver station (BTS) and software upgrades to the rest of the network.

In terms of mobile devices, almost all of the IS-2000 releases require new handsets. For example, the Release 0 handset cannot take advantage of the 1xEV-DV capabilities, but a Release D handset can take advantage of the Release 0 network. The different releases will require new handsets with new application-specific integrated circuits (ASICs).

Figure 3.31 shows a typical CDMA2000/IS-2000 system architecture, with its new elements as well as the 2G elements. The elements new to the CDMA2000 technology compared to 2G technologies are as follows:

• *AAA:* The authentication, authorization, and accounting server handles subscribers' requests for access to the mobile network.
• *PDSN:* The packet data serving node is a new element that performs packet-switched routing functions for all packet data sessions.

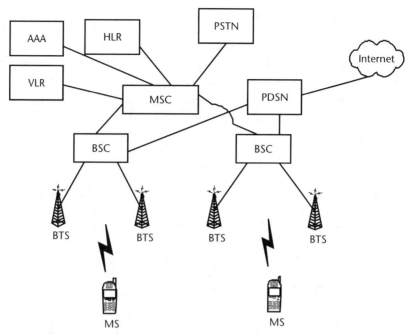

Figure 3.31   IS-2000 network architecture.

3.5.2   IS-856 Architecture

The IS-856 network architecture is very similar to the IS-2000 network architecture. The major difference is that IS-856 does not require MSCs since it does not support traditional voice service. The IS-856 architecture is shown in Figure 3.32, which illustrates the IS-2000 BTS with IS-856 channel modules as well as the separate BSCs.

3.6   Futuristic View of CDMA2000

This section highlights some of the contributions that are under review by 3GPP2 for the evolution of CDMA2000 technology beyond 1xEV-DO/1xEV-DV. It presents some potential new radio access technologies for CDMA2000-based systems. Table 3.7 presents some key attributes of the proposals submitted by several CDMA operators and manufacturers during the March 2005 3GPP2 meeting. The proposals are currently under review by the Technical Specification Group-CDMA2000 (TSG-C) committee [18]. At a high level the proposals lined up the evolution in terms of a short-term phase (2006–2007) and a long-term phase (2008 and beyond). The short-term phase focuses mainly on the multicarrier high rate packet data (HRPD) (that is, 1xEV-DO), while keeping the existing HRPD Revision A physical layer. The long-term phase combines the multicarrier approach with technologies such as interference cancellation, OFDM, and multiple input, multiple output (MIMO). Thus, the *projections* in Table 3.7 are very preliminary in nature, and readers should not consider them as final.

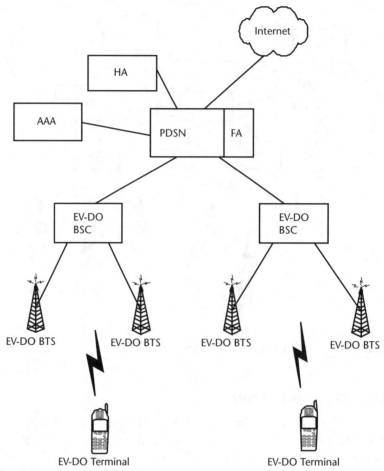

Figure 3.32 IS-856 network architecture.

3.7 CDMA2000 Technologies Comparison

In the preceding sections we described several technologies of the CDMA2000 family of standards. Here, Table 3.8 provides a worthwhile look at a high-level comparison among them.

3.8 Summary

In summary, the chapter provided details of the CDMA2000 1X technology and its migration path. The chapter illustrated the 1xRTT, 1xEV-DO, and 1xEV-DV technologies and provides information on the physical, link, and upper layers. The network topologies of 3G1x and EV-DO were also described. A futuristic view of CDMA2000 technology was presented and we also provided a comparison among several CDMA2000 technologies.

Table 3.7 Future CDMA2000 System's Expected Technical Attributes

Attribute	Short Term	Long Term
Air interface	Existing (1xEV-DO)	CDMA-OFDM mix
Multicarrier HRPD	Yes; up to 10 MHz	Up to 20 MHz
Peak data rates	10–15 Mbps (uplink and downlink)	75–100 Mbps (downlink); 25–50 Mbps (uplink)
Average data rates	2–3 Mbps (downlink); 1.5–2 Mbps (uplink)	25–30 Mbps (downlink); 5–10 Mbps (uplink)
Sector throughput	4–6 Mbps (downlink); 1–1.2 Mbps (uplink)	40–50 Mbps (downlink); 3–5 Mbps (uplink)
Latency	No change	Reduce latency
MIMO	No	Yes
ASIC	New	New
Backward compatibility	Yes	Minimal
Standard specification completion	May 2006	April 2007

Table 3.8 Comparison of CDMA2000 Technologies

Attribute	CDMA2000 1X	1xEV-DV Release D	1xEV-DO Revision A
FDD or TDD (in terms of spectrum)	FDD	FDD	FDD
Channel bandwidth	1.25 MHz, shared by voice and data	1.25 MHz, shared by voice and data	1.25 MHz, shared by voice and data
Voice	Supports circuit-switched voice, no VoIP support	Supports circuit-switched voice and VoIP	Supports VoIP
Peak data rates	Downlink: 153.6 Kbps Uplink: 153.6 Kbps	Downlink: 3.1 Mbps Uplink: 1.8 Mbps	Downlink: 3.1 Mbps Uplink: 1.8 Mbps
Average data rates	Downlink: 50–70 Kbps Uplink: 50–70 Kbps	Downlink: 450–650 Kbps Uplink: 300–500 Kbps	Downlink: 450–650 Kbps Uplink: 300–500 Kbps
Downlink sector throughput	200 Kbps	1,000 Kbps	1,000 Kbps
Uplink sector throughput	250 Kbps	700 Kbps	700 Kbps
Average data call setup time	5 seconds	<1 second	<1 second
Average one-way latency	150 ms	100 ms	100 ms
Fast call setup (<1 second)	No	Yes	Yes
Simultaneous voice and data	No	Yes	Yes with VoIP
Spectral efficiency	Downlink: 0.16 bps/Hz Uplink: 0.20 bps/Hz	Downlink: 0.8 bps/Hz Uplink: 0.56 bps/Hz	Downlink: 0.8 bps/Hz Uplink: 0.56 bps/Hz
Quality of service	No	Yes	Yes
Broadcast multicast	No	Yes	Yes
Support for CDMA2000 1X	—	Yes	Yes with hybrid terminals

References

[1]   "Introduction to CDMA2000 Standards for Spread Spectrum System," TIA/EIA/IS-2000-1, Standards & Technology Department, TIA, August 1999.

[2]   "Introduction to CDMA2000 Standards for Spread Spectrum System," TIA/EIA/IS-2000-1, Standards & Technology Department, TIA, June 2000.

[3]   "Physical Layer Standard for CDMA2000 Spread Spectrum System," TIA/EIA/IS-2000-2, Standards & Technology Department, TIA, August 1999.

[4]   "Physical Layer Standard for CDMA2000 Spread Spectrum System—Addendum 2," TIA/EIA/IS-2000.2-A-2, Standards & Technology Department, TIA, April 2002.

[5]   "Media Access Control (MAC) for CDMA2000 Spread Spectrum System," TIA/EIA/IS-2000-3, Standards & Technology Department, TIA, August 1999.

[6]   Willie, W. L., *Broadband Wireless Mobile*, New York: John Wiley & Sons, 2002.

[7]   "Signaling Link Access Control (LAC) for CDMA2000 Spread Spectrum System," TIA/EIA/IS-2000-4, Standards & Technology Department, TIA, July 1999.

[8]   "Upper Layer Signaling Standard for CDMA2000 Spread Spectrum System," TIA/EIA/IS-2000-5, Standards & Technology Department, TIA, July 1999.

[9]   Smith, C., and J. Meyer, *3G Wireless with WiMax and Wi-Fi: 802.16 and 802.11*, New York: McGraw-Hill, 2004.

[10] Yang, S., *CDMA RF System Engineering*, Norwood, MA: Artech House, 1998.

[11] "CDMA2000 High Rate Packet Data Air Interface Specification," TIA/EIA/IS-856, Standards & Technology Department, TIA, November 2000.

[12] "CDMA2000 High Rate Packet Data Air Interface Specification," TIA-856-A, Standards & Technology Department, TIA, April 2004.

[13] "Physical Layer Standard for CDMA2000 Spread Spectrum System—Release C," TIA/EIA/IS-2000.2-C, Standards & Technology Department, TIA, May 2002.

[14] "Introduction to CDMA2000 Spread Spectrum System," TIA-2000.1-D, Standards & Technology Department, TIA, March 2004.

[15] Sindhushayana, N. and P. Black, "Forward Link Coding and Modulation for CDMA2000 1xEV-DO (IS-856)," *Proc. 13th IEEE Int. Symp. Personal, Indoor and Mobile Radio Communications*, September 15–18, 2002, pp. 1839–1846.

[16] Esteves, E., "The High Data Rate Evolution of the CDMA2000 Cellular System," *Multiaccess, Mobility and Teletraffic for Wireless Communications*, Vol. 5, 2000, pp. 61–72.

[17] Esteves, E., "On the Reverse Link Capacity of CDMA2000 High Rate Packet Data Systems," *Proc. IEEE Int. Conf. on Communications*, April 28–May 2, 2002, pp. 1823–1828.

[18] "Draft Summary of TSG-C Air Interface Evolution Technical Experts Meeting (AIE)," 3GPP2 Meeting, March 10–11, 2005.

CHAPTER 4

Introduction to UMTS

The Third Generation Partnership Project was created by several standard development organizations to define the evolution of GSM systems. It came up with a radio access technology called wideband code-division multiple access (WCDMA) that meets the 3G wireless systems requirements set by the ITU. WCDMA is mainly used in Europe in the context of migration from GSM to the Universal Mobile Telecommunication System (UMTS). UMTS refers to the interconnection of a new type of radio access network, the UMTS Terrestrial Radio Access Network (UTRAN), to the adapted pre–Release 99 GSM/GPRS core network infrastructure. Release 99 was the first 3GPP release to introduce WCDMA and it was functionally frozen in December 1999. In addition to Release 99, 3GPP has also published additional releases that build on the evolution of the GSM/GPRS systems described in this chapter.

This chapter covers three main areas: the WCDMA radio air interface, features and architectures associated with additional releases of 3GPP, and high-speed downlink packet access (HSDPA), which is the next phase in the evolution of UMTS-based networks.

4.1 WCDMA Radio Interface

WCDMA radio technology is the backbone of the UMTS architecture, which includes many of the basic subsystems of GSM. However, some existing subsystems have been upgraded and new subsystems and functionalities have been added. The UMTS system has three basic parts (Figure 4.1): user equipment (UE), UTRAN, and the core network (CN). The interface defined between the UE and the UTRAN is the radio interface (Uu), and the interface between the core network and the UTRAN is called Iu [1].

The WCDMA radio interface is characterized by its protocols, for which it can be defined by two main groupings according to the end service: the user plane (U-plane) protocols and the control plane (C-plane) protocols. The former carries user data through the access stratum, whereas the latter controls the connections between the UE, the network, and the radio access bearers.[1] A functional layering of UMTS has also been agreed to by 3GPP that introduces the concept of access and nonaccess stratum (NAS). The access stratum offers services through the service access points (SAPs) to the NAS, which is highlighted in this section.

---

1.   The service provided by layer 2 is referred as the *radio bearer*.

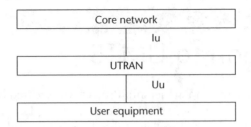

Figure 4.1 UMTS architecture.

The protocol architecture (Figure 4.2) splits the radio interface into three layers: the physical layer, the data link layer, and the network layer. The data link layer (layer 2 or L2) consists of medium access control (MAC), radio link control (RLC), Packet Data Convergence Protocol (PDCP), and broadcast/multicast control (BMC) sublayers. The network layer (layer 3 or L3) is further divided into radio resource control (RRC) and duplication avoidance. This hierarchical stratification provides a complete vision of the radio interface, from the functionality associated with each of the structured layers to the protocols that flow among them. Each lower layer is characterized by the services provided to the higher layers or entities. The physical layer offers information transfer services over its WCDMA radio medium for the upper layers. The MAC provides data transfer services for RLC and so on.

Each block in Figure 4.2 represents an instance of the respective protocol. SAPs for peer-to-peer communication are marked with circles at the interface between sublayers. The SAP between MAC and the physical layer provides the transport channels. The transport channels are defined by one or more transport formats, each of which is defined by encoding, interleaving, rate matching,[2] and mapping onto the physical channel. The structures of the logical, transport, and physical channels are shown in Figure 4.3. The SAPs between the RLC and the MAC sublayer provide the logical channels. The RLC layer provides specific types of SAPs for different RLC operation modes. The PDCP and BMC are accessed by PDCP and BMC SAPs, respectively. In the C-plane, the interface between "duplication avoidance" and higher L3 sublayers [call control (CC), mobility management(MM)] is defined by the general control (GC), notification (Nt), and dedicated control (DC) SAPs. The higher L3 functions such as CC and MM are assumed to belong to the NAS and therefore they are not described in this chapter [1–4].

4.1.1 Physical Layer

The physical layer (layer 1 or L1) offers information transfer services to the MAC and higher layers. Also, physical layer services are accessed through the use of transport channels. The physical layer provides multiple functions, including but not limited to error detection on transport channels, forward error correction encoding/decoding, interleaving/deinterleaving of transport channels, and modulation and spreading/demodulation and despreading of physical channels. The physical layer or layer 1 access scheme is based on wideband DS-CDMA technology with two duplex modes: FDD and TDD.

---

2. Rate matching repeats or punctures bits in order to match the physical channel bit rate.

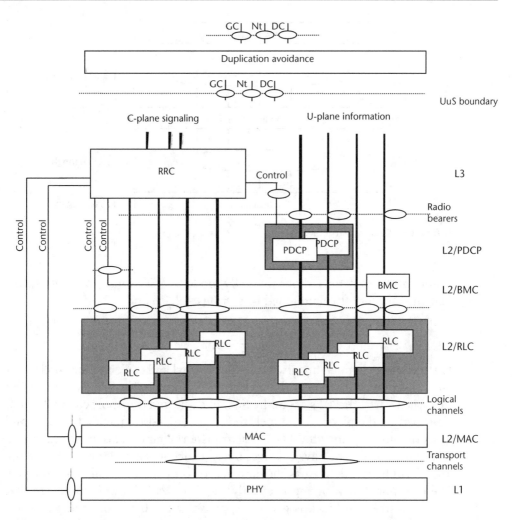

Figure 4.2 WCDMA radio interface protocol architecture [1]. Reprinted with permission from ETSI on behalf of the Third Generation Partnership Project (3GPP™). 3GPP™ TSs and TRs are property of ARIB, ATIS, ETSI, CCSA, TTA, and TTC who jointly own the copyright in them. They are subject to further modifications and are therefore provided to you "as is" for information purposes only. Further use is strictly prohibited.

- In the FDD mode, the uplink and downlink are transmitted on separate frequencies, and a physical channel is characterized by the code and frequency.
- In the TDD mode, the uplink and downlink are transmitted on the same carrier frequency and multiplexed in time. The physical channel is characterized by the code and timeslot and the chiprate is 3.84 Mcps, but a fixed chiprate does not imply a fixed service bit rate. Different symbol rates can be specified for each physical channel applying different spreading factors (SF)³ to each symbol.

The FDD mode [5–8] is based on CDMA, whereas the TDD mode [9–11] includes an additional TDMA component in addition to DS-CDMA. This chapter

---

3. The spreading factor is defined as the number of chips per data symbol.

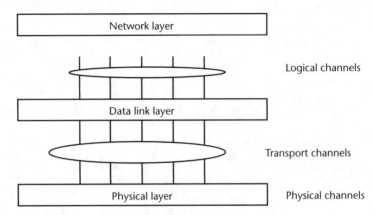

Figure 4.3   Channel structure.

mainly provides details about the FDD mode (see Sections 4.2.1.3 to 4.2.1.5); the TDD mode is briefly highlighted in Sections 4.2.1.6 and 4.2.1.8.

4.1.1.1   Transport Channels

The transport channels are always unidirectional and can either be common or dedicated. In common transport channels there is a need for in-band identification of the UEs when particular UEs are addressed. In dedicated transport channels the UEs are identified by the physical channel, that is, code and frequency for FDD and code, timeslot and frequency for TDD. The transport channels are used in both the FDD and TDD modes. The physical layer operates according to L1 radio frame timing. A transport block is defined as the data accepted by the physical layer to be encoded. The transport block timing is exactly the same as L1 frame timing; for example, every transmission block is generated precisely in 10 ms, or multiple of 10 ms. A 10-ms frame (Figure 4.4) is divided into 15 slots (2,560 chips/slot at a chiprate of 3.84 Mcps).

A mobile can set up multiple transport channels simultaneously, each having its own transport characteristics (e.g., offering different error correction capabilities). Each transport channel can be used for information stream transfer of one radio bearer or for L2 and higher layer signaling messages. The multiplexing of the transport channel onto the same or different physical channel is carried out by L1.

4.1.1.2   Physical Channels

There are two types of physical channels—dedicated and common—each with its own specific purpose. They are designed so they cycle through a prescheduled

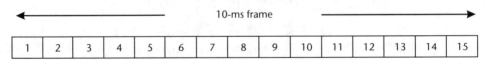

Figure 4.4   WCDMA frame structure [12].

sequence of operations, and different types of information are transmitted on each timeslot during this cycle. The following are the physical channels:

- *Primary common control physical channel (PCCPCH):* A downlink channel that continuously transmits the system's identification and access control information.

- *Secondary common control physical channel (SCCPCH):* A downlink channel that carries forward access channel (FACH) information along with paging channel (PACH) messages for mobile telephones that have registered with the system.

- *Physical random access channel (PRACH):* An uplink channel that allows a mobile telephone to randomly transmit access requests when a handset attempts to access the cellular system.

- *Dedicated physical data channel (DPDCH):* Transfers user data on the uplink and downlink channels. It transmits data generated at L2 and above (i.e., dedicated and traffic channels).

- *Dedicated physical control channel (DPCCH):* Transfers control information between the system and the handset.

- *Physical downlink shared channel (PDSCH):* Provides control information to the mobiles operating within the coverage area.

- *Physical common packet channel (PCPCH):* Designed to carry packet data. Its operation is similar to the RACH channel.

- *Synchronization channel (SCH):* Assists in the cell searching process and is divided into primary and secondary channels. The primary channel uses a 256 sequence that is identical for every cell in the network. The secondary SCH is chosen from a set of 15 different codes, 256 chips in length. The secondary SCH identifies the scrambling sequence for the downlink channels. The primary SCH and secondary SCH are transmitted in parallel.

- *Common pilot channel (CPICH):* An unmodulated code channel that is scrambled with the cell-specific primary scrambling code. It is used for channel estimation at the UE for the dedicated channel and provides the channel estimation reference for the common channels when they are not associated with the dedicated channels. There are two types of pilot channels: primary and secondary. There is only one primary channel code per cell or sector. The secondary pilot channel may have other codes.

- *Acquisition indication channel (AICH):* Used to assign a mobile to a data channel (DCH), where it can begin to communicate with the system.

- *Page indication channel (PICH):* A fixed rate (SF = 256) channel used to carry the page indicator. It is always associated with an SCCPCH, to which a PCH transport channel is mapped.

The mapping of transport channels onto physical channels is shown in Table 4.1.

Table 4.1 Mapping of Transport
Channels to Physical Channels

Transport Channel	Physical Channel
—	CPICH
BCH	CCPCH
PACH	CCPCH
PCH	CCPCH
RACH	PRACH
CPCH	PCPCH
DCH	DPDCH
DCH	DPCCH
DSCH	PDSCH
—	PICH
—	SCH
—	AICH

4.1.1.3 FDD Mode Multiplexing and Coding

The multiplexing and channel coding functionality is a combination of error detection, error correction, rate matching, interleaving, and multiplexing. The information is processed in transmission time intervals of 10, 20, 40, or 80 ms, in basic processing units called *transport blocks*. The step information follows until it is ready to be sent through the channel. It can be clearly differentiated into two main parts. The first step sets the information with the corresponding transport format for each transport channel, which simply means that the information is coded and arranged in order to perform the next step in the process, which is multiplexing. The multiplexing is done over different transport channels. A code-composite transport channel (CCTCH) is generated, and this would be mapped to one or more different physical channels [2].

The steps set the transport format differ when referring to the uplink or downlink. In the downlink, the transmission does not necessarily have to be continuous; discontinuous transmission bits can be inserted to determine in which time intervals there is no transmission. In the uplink, rate matching is accomplished after first interleaving and radio frame segmentation have been performed. But there are also three steps in the process that are common for both: CRC, transport block concatenation/code block segmentation, and channel coding.

In the uplink, the steps that occur after the initial three steps are radio frame equalization, first bit interleaving, radio frame segmentation, and rate matching. In the downlink, after channel coding, rate matching, first insertion of discontinuous transmission (DTX) indication bits, first bit interleaving and radio frame segmentation are performed. There are two main differences between the uplink and downlink. In the multiplexing part, the steps involved are transport channel multiplexing, second insertion of DTX indication bits (only in the downlink), physical channel segmentation, second bit interleaving, and physical channel mapping.

In the CRC attachment, a number of bits, specified by higher layers are inserted in order to provide the transport block with a redundancy check. The number of bits inserted belongs to the set {0, 8, 12, 16, and 24} and is obtained through four different cyclic generator polynomials, respectively. Once the CRC is performed, the serial concatenation of all the transport blocks to be processed might not give a total length adequate for the next step, which is coding. The channel coding defines a maximum size for the block to be coded (code block), depending on the coding that would be used. There are three possibilities:

- Convolution coding (CC) with a maximum code block of 504 bits;
- Turbo coding (TC) with a maximum code block of 5,114 bits;
- No coding.

If the total number of bits after concatenation exceeds the corresponding maximum code block length, segmentation will be performed to obtain a number of code blocks of equal length. The transport channel that is processed determines the possible coding scheme(s) and the coding rate(s) (Table 4.2). Convolutional coding for coding rates of 1/2 and 1/3 has a memory order of 9. The generator polynomials are 561 (octal) and 753 (octal), respectively, for a coding rate of 1/2 and 557(octal), 663 (octal), and 711 (octal) for a coding rate of 1/3.

Radio frame equalization is needed only in the uplink in order to set the number of bits at the input to a multiple of the number of radio frames (10 ms) in the transmission time interval. This operation is not necessary in the downlink because the output of the rate matching that is the function performed following the channel coding is already a multiple of the number of radio frames. At this point in both the uplink and downlink, the number of bits in the radio frame is already fixed; considering that the radio frame transmission time interval is 10 ms, the number of bits in the radio frame is fixed by upper layer signaling with the spreading factor. First bit interleaving is performed while setting the transport format, which is block interleaving with the intercolumn permutation. The number of columns and the permutation between them is fixed and depends on whether the transmission time interval is 10, 20, 40, or 80 ms. The radio frame segmentation is only necessary if the transmission time interval is longer than 10 ms, in which case the input sequence is divided and mapped into consecutive radio frames. Each 10-ms processed radio frame from each of the transport channels that will generate the CCTCH is

Table 4.2 Coding Schemes for Transport Channels in FDD [2]

Transport Channel	Coding Scheme			Coding Rate	
	CC	TC	No Coding	CC	TC
BCH	X	—	—	1/2	—
PCH	X	—	—	1/2	—
RACH	X	—	—	1/2	—
FACH	X	X	X	1/2, 1/3	1/3
CPCH	X	X	X	1/2, 1/3	1/3
DSCH	X	X	X	1/2, 1/3	1/3
DCH	X	X	X	1/2, 1/3	1/3

delivered to the multiplexing unit. The number of transport channels that can be multiplexed depends on the nature of the transport channel. The same thing happens with the number of physical channels that can be generated from the CCTCH after the physical channel segmentation. The second interleaving is made for each of the physical channels. It is block interleaving with intercolumn permutation that randomizes the bit position of the different transport channels. The last step is the physical channel mapping where the outputs bits of the second interleaver are mapped to the different physical channels. Each physical channel is spread and modulated according to its specific characteristics, as described in the next section [2].

4.1.1.4   FDD Mode Spreading and Modulation

In the FDD mode, the spreading and modulation processes have been defined separately for the uplink and downlink communications.

Uplink and Downlink Spreading
Spreading is applied to the physical channels and is divided into two parts. The first is channelization operation, in which the bit sequence is spread with orthogonal sequences that preserve the orthogonality between sequences even when the spreading factor is different. The second operation is scrambling, in which a scrambling code is applied to the spread signal. Scrambling is applied on top of channelization, as shown in Figure 4.5.

During channelization, the data symbols are transformed into a number of chips, thus increasing the bandwidth of the signal. This channelization is conducted using orthogonal variable spreading factor (OVSF) codes. In the uplink, OVSF codes preserve the orthogonality among different users' channels and in the downlink between the different channels in the base station. The code allocation in the uplink is fixed and depends on the SF applied to that particular channel. In the downlink the codes for the PCCPH and CPICH are fixed, whereas the codes for the rest of the physical channels are assigned by UTRAN. The availability of the number of codes depends on the spreading factor and is defined using the code tree of Figure 4.6. In Figure 4.6, the channelization codes are uniquely described as $C_{SF,k}$, where SF is the spreading factor of the code and k is the code number, $0 <= k <= SF - 1$. Each level in the code tree defines channelization codes of length SF, corresponding to a spreading factor of SF in Figure 4.6.

In a WCDMA scheme, all users transmit on the same frequency and are distinguished by their unique scrambling codes. The receiver correlates the received signal with a synchronously generated replica of the scrambling code to recover the

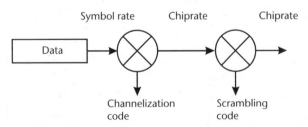

Figure 4.5   Relationship between channelization and scrambling [13].

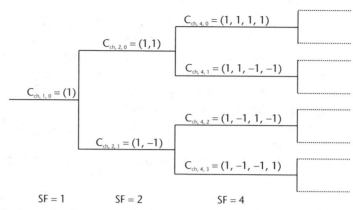

Figure 4.6   Code tree for the generation of OVSF codes [5].

original information-bearing signal. Scrambling doesn't change the bandwidth of the signal, but only makes the signal from different users separable from each other.

The scrambling codes are different for the uplink and downlink. In WCDMA uplink transmissions, the scrambling codes can either be short and long. The long codes are essentially Gold codes, while short codes are VL-Kasami codes. There are 2^{24} possible long scrambling sequences, and these are assigned by higher layers. The different sequences are generated through different initialization of the 25 memory position shift registers that generate the sequence; the repetition period of each sequence is every 2^{24} bits. Thus, 25-degree generator polynomials are truncated to the 10-ms frame length in 38,400 chips at the rate of 3.84 Mcps. For short sequences the number of possible sequences is the same but they are generated through three shift registers of eight memory positions; consequently, the repetition period is reduced to 256 (2^{8}) chips. This means that within a 10-ms frame of 38,400 chips, 150 periods of the code are present [14].

In the downlink, Gold codes of length 2^{18} are used, but they are truncated to form a cycle of 10-ms frames. These scrambling codes are generated by two shift registers of 18 memory positions, but not all of them are used. From the whole set of codes, a subset of 8,192 codes is used that is divided into primary and secondary scrambling codes. There are 512 primary codes and 15 secondary scrambling codes. One-to-one mapping is used between each primary and 15 secondary codes in a set such that the ith primary scrambling corresponds to the ith set of the scrambling codes. Thus, scrambling codes $m = 0, 1, ..., 8,191$ are used. The primary codes are divided into 32 code groups with 16 codes in each group to facilitate a fast cell search procedure in the next section. This code structure helps to maintain only one primary code per cell. The primary CCPCH, primary CPICH, PICH, AICH, AP-AICH, CD/CA-ICH, CSICH, and S-CCPCH carrying PCH are always transmitted using the primary scrambling code. The other downlink physical channels can be transmitted with either the primary scrambling code or a secondary scrambling code from the set associated with the primary scrambling code of the cell [2–5, 14].

Uplink and Downlink Modulation
After spreading, the next step is to perform modulation. The data modulation scheme used in both uplink and downlink is QPSK. However, how the QPSK

symbols are generated and from what information sources differ with the type of logical channel being mapped and if we are working in the uplink or the downlink. In the uplink, for example, different data channels and a control-associated channel can be mapped into just one physical channel by means of the channelization sequences that would separate the different channels and the orthogonality provided by the phase (I) and quadrature (Q) branches of the QPSK signal. In the downlink, data modulation is QPSK where each pair of bits is serial-to-parallel converted and mapped to the I and Q branches, respectively. The I and Q branches are then spread to the chiprate with the channelization code and then scrambled by the scrambling code.

4.1.1.5   Physical Layer Procedures

This section discusses four different physical layer procedures in the FDD mode of the UTRAN.

- *Cell search*: During the initial cell search, the UE first searches for the strongest base station. In the second step, the UE uses the secondary SCH to find frame synchronization and identify the code group of the cell found in the first step. Finally, the UE determines the scrambling code used by the found cell [15].
- *Power control*: WCDMA employs closed-loop power control in both the uplink and downlink. The basic power control rate is 1,600 Hz, and the power controlled step can be varied adaptively according to the UE speed and the operating environment. The signal-to-interference ratio (SIR) is the basic parameter used for the power control mechanism. The target SIR values are controlled by an outer-loop power control. This outer loop measures the link quality, which is typically a combination of frame and bit error rates (BERs), depending on the service, and adjusts the SIR targets accordingly. It ensures that the lowest possible SIR target is used at all times to maximize network capacity. WCDMA also uses open-loop power control to adjust the transmitting power of the physical random access channel (RACH).
- *RACH procedure*: The RACH is typically used for signaling purposes, to register the UE with the network after power-on, to perform a location update after moving from one location to another, or to initiate a call.
- *Transmit diversity procedure*: The downlink can use either an open-loop or closed-loop transmit diversity procedure to improve link performance. In the open loop, there is no information from the uplink, but information is sent from two antennas in the downlink direction. In the closed loop, the base station has information provided by the UE, which is included in one of the control fields of the uplink dedicated control channel. The UE computes this information through measurements in CPICH of the estimate channels seen from each antenna.

4.1.1.6   TDD Channel Structure

The TDD mode has a transport channel structure similar to the one in the FDD mode. However, the TDD mode has an additional transport uplink shared access channel that is shared among several UEs to carry dedicated control or traffic data.

All of the physical channels have a three-layer structure based on system frame numbering, radio frames, and timeslots. Depending on the resource allocation, the configuration of radio frames or timeslots becomes different. All physical channels need guard symbols in every timeslot. The timeslots such as a TDMA component are used to separate different user signals in the time and the code domain. The physical channel signal format is shown in Figure 4.7. A physical channel is a TDD burst, which is transmitted in a particular timeslot within allocated radio frames. The allocation can be continuous (i.e., the timeslot in every frame is allocated to the physical channel) or discontinuous (i.e., the timeslot in a subset of all frames is allocated only). The TDMA frame has a duration of 10 ms and is subdivided into 15 timeslots (TS) of $2,560 \times T_c$ duration each. A timeslot corresponds to 2,560 chips. Each frame can either be allocated in the uplink or downlink. With such flexibility, the TDD mode can be adapted to different environments and deployment scenarios.

4.1.1.7 TDD Multiplexing, Channel Coding, and Interleaving

The data stream going to or coming from MAC and higher layers (transport block/transport block set) is encoded/decoded to offer transport services over the radio transmission link. The channel coding process is similar to the one described in the FDD (see section 4.2.1.5). Details of TDD multiplexing and channel coding can be found in [6]. At a high level, the channel coding scheme is a combination of error detection, error correction, rate matching, interleaving, and mapping of transport channels onto physical channels. Four different coding schemes are available:

- Rate $R = 1/3$, $K = 9$, convolutional coding (CC), where K is the constraint length;
- Rate $R = 1/2$, $K = 9$, convolutional coding (CC), where K is again the constraint length;
- Rate $R = 1/3$ or $1/2$, $K = 3$ or 4, turbo codes, where K is the constraint length of the recursive systematic convolutional encoder;
- No coding.

4.1.1.8 TDD Spreading and Modulation

The information rate of the channel varies, with the symbol rate being derived from the 3.84-Mcps chiprate and the spreading factor. Spreading factors range from 16

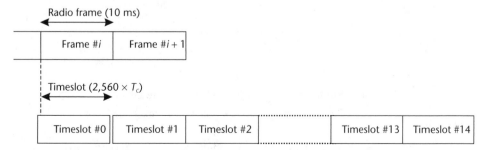

Figure 4.7 Frame structure [16].

to 1 for TDD uplink and downlink. Thus, the QPSK modulation symbol rates vary from 3.84 megasymbols per second (Msps) to 240 symbols/s. Like FDD, TDD also uses a QPSK modulation scheme for data and the same chiprate of 3.84 Mcps.

4.1.1.9   FDD Versus TDD

Table 4.3 illustrates the main WCDMA air interface parameters in both the FDD and TDD modes. This table shows that both FDD and TDD operate very similarly; however, there are some differences in transmitter power control and synchronization processes.

4.1.2   Data Link Layer

As stated at the beginning of this chapter, the data link layer is subdivided into the following sublayers: MAC, RLC, PDCP, and BMC. This section briefly describes these sublayers.

4.1.2.1   Medium Access Control Sublayer

The MAC sublayer in general provides data transfer services for the RLC sublayer. The MAC sublayer also controls the multiplexing of data streams originating from different services, multiplexing/demultiplexing of upper layer PDUs, and priority handling of data flows for one UE and between UEs [1, 17].

The MAC sublayer provides data transfer services on logical channels. The configuration of logical channel types is depicted in Figure 4.8. There are two types of logical channels: control and traffic. The control channels are used for transfer of control plane information only, and the traffic channels are used for the transfer of user plane information only.

Table 4.3   WCMDA Air Interface Main Parameters

	FDD Mode	TDD Mode
Access method	DS-CDMA	
Duplexing mode	FDD	TDD
Chiprate	3.84 Mcps	
Frame length	10 ms	
Slot length	0.667 ms	
Multirate concept	Uplink: variable SF and/or multicode Downlink: variable SF and/or multicode	
Channel coding	Convolutional code (R = 1/3 or 1/2, K = 9) or turbo codes (R = 1/3 or 1/2, K = 3)	
Modulation	Data: QPSK; spread: QPSK	
Power control	Downlink and uplink: closed loop Uplink: open loop	
Synchronization	Asynchronous	Synchronous

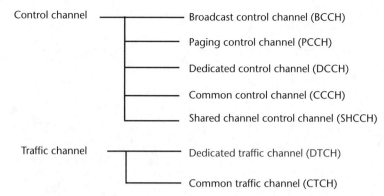

Figure 4.8 Logical channel structure [1].

Figure 4.9 shows the mapping between logical channels and transport channels. The MAC sublayer controls radio resources on a "fast basis." Given the transport format combination set assigned by RRC, MAC selects the appropriate transport format within an assigned transport format set for each active transport channel depending on the source rate and total interference threshold level.

MAC Architecture
The MAC architecture is constructed from the three MAC entities or peer entities that handle several transport channels:

- *MAC-b:* The MAC entity that handles the broadcast transport channel.
- *MAC-c/sh:* The MAC entity that handles the following transport channels:
  - PCH, FACH, RAH, uplink common packet channel (UL CPCH) (note that the CPCH exists only in FDD mode), DSCH, and USCH (note that the USCH exists only in TDD mode).
- *MAC-d:* The MAC entity that handles the DCH transport channels.

The exact functions completed by these entities are different in the UE from those completed in the UTRAN. The details of these exact functions can be found in [17].

MAC Services to Upper Layers
This section highlights the services provided by the MAC sublayer to the upper layers. These three key services are as follows [1]:

- *Data transfer:* This service provides unacknowledged transfer of MAC PDUs among peer MAC entities. This service does not provide any data segmentation. Therefore, the segmentation/reassemble function should be achieved by an upper layer.
- *Reallocation of radio resources and MAC parameters:* This service performs on request of the RRC execution of radio resource reallocation and change of MAC parameters (i.e., reconfiguration of MAC functions such as change of identity of UE, change of transport format combination set, and change

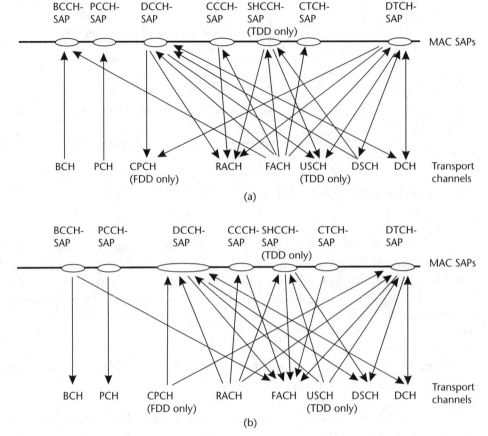

Figure 4.9 Logical channels mapped onto transport channels as seen (a) from the UE side and (b) from the UTRAN side [1].

of transport channel type). Additionally, the MAC can handle resource allocation autonomously in the TDD mode.

- *Reporting of measurements:* The MAC sublayer reports local measurements such as traffic volume and quality indication to RRC.

4.1.2.2 Radio Link Control Sublayer

The RLC sublayer is responsible for efficient transmission or retransmission under variable bit rates. Therefore, a minimal segmentation overhead, a simple retransmission protocol, and an optimized transmission or retransmission unit size are required for the RLC design in different radio environments (e.g., different fading scenarios). RLC is configured by L3 to provide different levels of QoS. This QoS is achieved by adjusting the maximum number of retransmissions according to service delay criteria. For nonreal-time traffic, RLC provides low-level selective retransmission-ARQ-functionality with CRC-based error detection. The RLC sublayer provides user data transfer, error correction, flow control, and so forth [1, 18].

RLC Architecture

Figure 4.10 gives an overview model of the RLC sublayer. The figure illustrates the different RLC entities, of which there are three types: transparent mode (TM), unacknowledged mode (UM), and acknowledged mode (AM).

There is one transmitting and one receiving RLC entity for each TM and UM service. There is one combined, transmitting, and receiving entity for the AM service. The word *transmitted* means "submitted to a lower layer" unless otherwise explicitly stated. The dashed lines between the AM entities show the possibility of sending the RLC PDUs on separate logical channels, for example, control PDUs on one channel and data PDUs on the other. Details about the TM, UM, and AM entities can be found in [18].

RLC Services to Upper Layers

Five key services are provided by the RLC to upper layers:

- *Transparent data transfer:* This service transmits higher layer PDUs without adding any protocol information, possibly including segmentation/reassembly functionality.
- *Unacknowledged data transfer:* This service transmits higher layer PDUs without guaranteeing delivery to the peer entity.
- *Acknowledged data transfer:* This service transmits higher layer PDUs and guarantees delivery to the peer entity. If the RLC is unable to deliver the data correctly, the user of the RLC at the transmitting side is notified. For this service, both in-sequence and out-of-sequence delivery are supported.
- *QoS maintenance:* The retransmission protocol is configurable by L3 to provide different levels of QoS, which can be controlled.
- *Notification of unrecoverable errors:* RLC notifies the higher layer about errors that cannot be resolved by the RLC itself by means of normal exception handling procedures, for example, by adjusting the maximum number of retransmissions according to delay requirements.

4.1.2.3   Packet Data Convergence Protocol Sublayer

As stated in the introductory section, PDCP provides transmission of higher PDUs in different RLC modes, mapping the network protocol into an RLC entity. This sublayer is defined for the packet-switched (PS) domain only, and it supports header compression and decompression of IP data streams [1, 19].

PDCP Architecture

The PDCP sublayer is configured by the upper layer through the PDCP-C-SAP (control service access point), as shown in Figure 4.11. This figure shows the model of the PDCP within the radio interface protocol architecture.

Every PS domain radio access bearer (RAB) is associated with one radio bearer (RB), which in turn is linked with one PDCP entity. Each PDCP entity that is located in the PDCP sublayer is mapped to either one AM RLC entity or one or two UM or TM RLC entities. When a PDCP entity is mapped to two UM or TM RLC entities, each RLC entity is used for a different direction. Every PDCP entity uses zero, one,

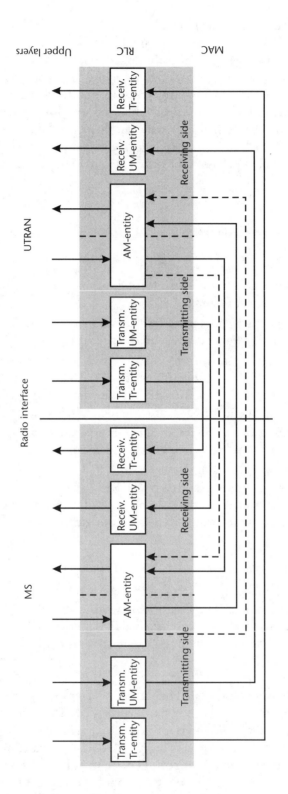

Figure 4.10 Overview model of the RLC sublayer [18].

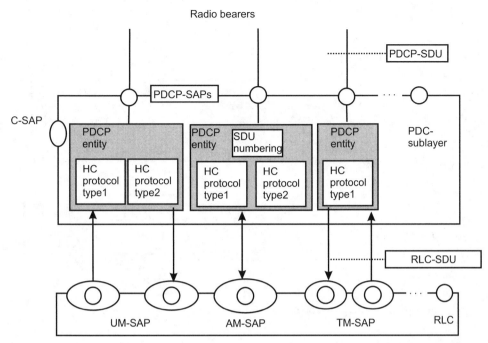

Figure 4.11 PDCP structure [19].

or several different header compression protocol types. Several PDCP entities may be defined for a UE with each using the same or a different protocol type.

PDCP Services to Upper Layers
The PDCP sublayer mainly provides the following two key functions to the upper layers:

- Transfer of user data;
- Maintenance of PDCP SDU sequence numbers.

4.1.2.4 Broadcast Multicast (BMC) Sublayer

The BMC is a sublayer of the data link layer that exists only in the user plane. It is located above the RLC sublayer. The L2/BMC sublayer is assumed to be transparent for all services except broadcast/multicast. It monitors traffic, schedules BMC messages, delivers broadcast messages to upper layers, and so forth [1, 20].

BMC Architecture
Figure 4.12 shows the model of the L2/BMC sublayer within the UTRAN radio interface protocol architecture. At the UTRAN side, the BMC sublayer consists of one BMC protocol entity per cell. Each BMC entity requires a single common traffic channel (CTCH), which is provided by the MAC sublayer through the RLC sublayer. The BMC uses the unacknowledged mode service of the RLC.

It is assumed that there is a function in the RLC above the BMC that resolves the geographical area information of the cell broadcast (CB) message (or, if applicable,

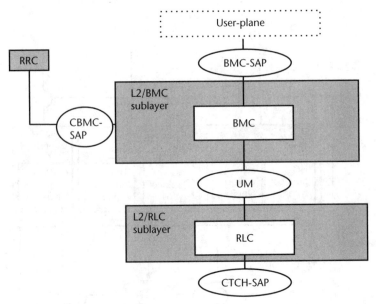

Figure 4.12 BMC protocol model [20].

performs evaluation of a cell list) received from the cell broadcast center (CBC). A BMC protocol entity serves only those messages at the BMC-SAP that are to be broadcast into a specified cell.

BMC Services
The BMC provides a broadcast/multicast transmission service in the user plane in the unacknowledged mode.

4.1.3 Network Layer

The network layer in UTRAN is divided into control and user planes. As stated earlier, in the C-plane, layer 3 is split into radio resource control and duplication avoidance. RRC interfaces with layer 2 and terminates in the UTRAN. Although the duplication avoidance sublayer terminates in the core network, it is part of the access stratum, and it provides the access stratum services to higher layers.

4.1.3.1 RRC Model

Figures 4.13 and 4.14 illustrate the RRC models for the UE and UTRAN sides, respectively. The functional entities that are part of this model are as follows [21]:

- Routing of higher layer messages to different MM/CM entities (UE side) or different core network domains (UTRAN side) is handled by the routing function entity (RFE).
- Broadcast functions are handled by the broadcast control function entity (BCFE). The BCFE is used to deliver the RRC required at the GC-SAP. The BCFE can use the lower layer services provided by the Tr-SAP and UM-SAP.

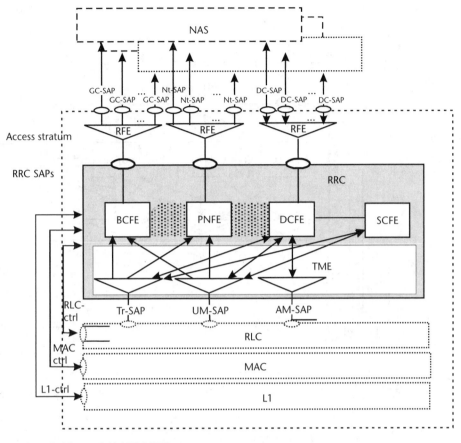

Figure 4.13 UE side model of RRC [21].

- Paging of UEs that do not have an RRC connection is controlled by the paging and notification control functional entity (PNFE). The PNFE is used to deliver the RRC services that are required at the Nt-SAP. The PNFE can use the lower layer services provided by the Tr-SAP and UM-SAP.
- The dedicated control function entity (DCFE) handles all functions specific to one UE. The DCFE is used to deliver the RRC services that are required at the DC-SAP. It can use lower layer services of the UM/AM-SAP and Tr-SAP depending on the message to be sent and on the current UE service state.
- In TDD mode, the DCFE is assisted by the shared control function entity (SCFE) located in the C-RNC that controls the allocation of both uplink and downlink shared physical channels via lower layer services of UM-SAP and Tr-SAP.
- The transfer mode entity (TME) handles the mapping between the different entities inside the RRC layer and the SAPs provided by RLC.

4.1.3.2 RRC Services

The RRC offers the following services to upper layers [21]:

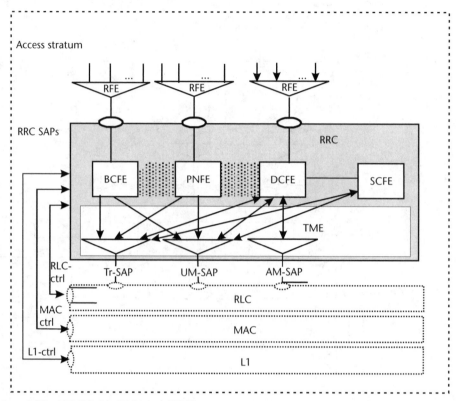

Figure 4.14 UTRAN side RRC Model [21].

- *General control:* The GC-SAP provides an information broadcast service. This service broadcasts information to all UEs in a certain geographical area.
- *Notification:* The Nt-SAP provides paging and notification broadcast services. Unlike GC-SAP, this service broadcasts in a certain geographical area, but the broadcasts are addressed to a specific UE(s).
- *Dedicated control:* The DC-SAP provides services for establishment/release of a connection and transfer of messages using this connection.

4.2   3GPP Network Architecture

The evolution of GSM has been standardized by 3GPP in several releases, starting with Release 1999 (R99) and moving forward through Release 4 (Rel-4), Release 5 (Rel-5), Release 6 (Rel-6), Release 7 (Rel-7), and Release 8 (Rel-8). The different releases either correspond to the different radio access technologies or include enhancements to the radio technology of the earlier release or both. For example, consider the following:

- *R99:* First WCDMA enabler with voice and data support; functionally frozen in December 1999.
- *Rel-4:* First 3GPP release that supports enhanced data rates for GSM evolution (EDGE) and TD-SCDMA; functionally frozen in March 2001.

- *Rel-5:* First 3GPP release that supports HSDPA, IP transport in radio access network, and IP multimedia subsystem (IMS); functionally frozen in June 2002.

- *Rel-6:* Release 6 is the first 3GPP release that supports high-speed uplink packet access (HSUPA). Rel-6 functional activities have been frozen.

- *Rel-7 and Rel-8:* Releases 7 and 8 enhance IMS, and introduce interworking between UMTS and wireless local access networks, long-term evolution of the 3G radio access network (Release 8), and so forth. These releases have not yet been frozen.

Thus, 3GPP evolved the GSM networks by taking a phased approach that is suitable for the fast changing world of wireless communications. In the next few sections, the network architectures corresponding to the R99, Rel-4, and Rel-5 releases are discussed [2, 22, 23].

4.2.1 Release 99

The first 3GPP release, R99, widely known as UMTS, evolved from the GSM and GPRS systems. The network architecture of UMTS (R99) is shown in Figure 4.15. In this figure, the dashed lines represent signaling links, and the solid lines represent data and signaling links. The mobile station (MS) is the physical equipment used by the users to access the two radio access networks (RANs) supported by the CN GSM base station subsystem (BSS) and UTRAN (UMTS terrestrial RAN).

As shown in Figure 4.15, the two RANs are connected to the core network via standardized interfaces. In particular, the GSM BSS is connected to the circuit-switched (CS) domain via the A interface (specified in [24, 25]) and to the PS

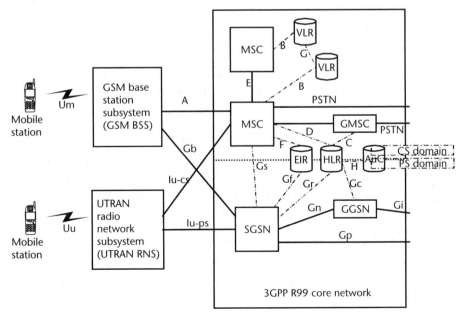

Figure 4.15 3GPP R99 network architecture [2].

domain via the Gb interface (specified in [26, 27]). On the other hand, UTRAN is connected to the CS domain via the Iu-cs interface and to the PS domain via the Iu-ps interface. Note that the signaling procedures in the control plane across the two logical interfaces are identical; that is, there is a common signaling protocol, called Radio Access Network Application Part (RANAP). However, the U-plane protocols are different in Iu-cs and Iu-ps. In addition, the two neighboring radio network controllers in the UTRAN are connected via Iur. UTRAN uses the WCDMA radio access technology, which supports both circuit-switched voice and packet data.

The R99 architecture inherits a lot from the GSM model on the core network side. The CN of R99 is an evolution of the GSM core network, which is based on mobile application part (MAP). The mobile switching center (MSC) has very similar functions both in GSM and UMTS. In addition to UTRAN, R99 also supports the legacy GSM BSS radio access network. This backward compatibility feature assists the legacy mobiles to operate in 3GPP R99 networks in a seamless fashion. In 3GPP R99, as well as in all later releases (Rel-4 through Rel-7), the core network is divided into two distinct domains: the CS domain and the PS domain. The CS domain is composed of the elements that provide services that require dedicated circuits, which is true of many GSM services. In the CS domain, the UMTS connects to the PSTN through the MSC. Also, connections originating from or terminating in the fixed networks are handled by a dedicated gateway MSC or GMSC. On the other hand, the PS domain is composed of the elements that utilize resources on a demand basis. In the PS service domain, UMTS connects to the packet data network (PDN) through the serving GPRS support node (SGSN) and gateway GPRS support node (GGSN). The SGSN in the PS domain is "equivalent" to the MSC in the CS domain. In general, packet switching is more efficient for packet data communications, whereas circuit switching is preferred for voice-type traffic. In addition to these key elements, several databases are available for call control and network management: the home location register (HLR), the visitor location register (VLR), and the authentication center (AUC) [28].

4.2.1.1 Release 99 Network Elements

The various components of the UTRAN and CN are briefly described in the next few sections, and details about the GSM BSS can be found in Chapter 2.

Mobile Station Elements
The MS uses the so-called Um radio interface to access the GSM BSS and the Uu interface to access the UTRAN. The radio interface Uu is between the MS and Node-Bs and is based on the WCDMA technology. Dual-mode (GSM/UMTS) mobile stations have the capability to use both Um and Uu interfaces. The MS comprises the mobile equipment (ME) and a smart card, formally called a Universal Integrated Circuit Card (UICC), that contains a SIM application and/or a UMTS subscriber identity module (USIM) application. When the MS contains a USIM application, it is also referred to as user equipment in some of the 3GPP specifications. More information about the MS functional split and UICC can be found in [29] and [30], respectively.

UTRAN Elements

The UTRAN has two key components: Node-B and the RNC. Several Node-Bs together are controlled by one RNC and those (Node-Bs plus one RNC) jointly form the radio network subsystem (RNS). The RNS is responsible for the radio resources and transmission/reception in a set of cells. A cell is formed by the radio coverage area of a Node-B [31]. As shown in Figure 4.15, the RNS is connected to the core network through the Iu interface; note that the reference point (interface) between two RNCs is referred to as Iur. Each RNC is assigned a pool of radio resources (e.g., frequencies or CDMA codes) and is responsible for managing those resources and allocating them to the mobile users on a demand basis.

Each UE connected to UTRAN is served by a specific RNC, which is called a serving RNC (SRNC). The SRNC controls the signaling connection between the UE and the UTRAN and also handles the Iu signaling connection for the same UE. The RNC that controls one or more Node-Bs serves as the controlling RNC (CRNC). There is only one CRNC for any Node-B and that CRNC has the overall control of the logical resources of that Node-B. The SRNC and CRNC may or may not be implemented in the same RNC. As mentioned earlier, the Iu interface is used to connect UTRAN to the core network. The Iu interface is logically divided into two interfaces: Iu-cs ad Iu-ps. The Iu-cs refers to the logical connection between the RAN and the CS domain, whereas the logical connection between the RAN and PS domain is Iu-ps.

GSM BSS

The BSS includes two types of elements: the base station transceiver station (BTS), which handles the radio interfaces toward the MS, and the base station controller (BSC), which manages the radio resources and controls handoffs. A BSC can manage several BTSs [31]. It has two interfaces: A interface and Gb interface. The A interface operates between an MSC and a BSS and is specified primarily in [25]. The signaling protocol on this interface is called the BSS application part (BSSAP), and it implements procedures for the provision of BSS management, call handling, and mobility management. The interface between the SGSN and BSS is the Gb interface, which supports the BSS GPRS Protocol (BSSGP) and provides flow control in the downlink direction.

Gateway GPRS Support Node

The 3G-GGSN is a network element in the PS domain that serves as a gateway providing connectivity to external packet data networks (PDNs) over the Gi interface. It could be considered as a typical router (home agent) implementing additional functionality for supporting mobile services. Such additional functionality includes the GPRS Tunneling Protocol (GTP), which manages user mobility by establishing and dynamically updating GTP tunnels within the PS domain. It may optionally communicate with the HLR via the Gc interface. It may provide OAM functionality, and it also collects charging information related to external networks usage by the user. The 2G-GGSN and 3G-GGSN, which are used in GSM/GPRS and UMTS systems, respectively, have very similar functionality, except that the evolution from 2G-GGSN to 3G-GGSN requires minor upgrades to a few of the protocols.

Serving GPRS Support Node

The 3G-SGSN is one of the main core network elements for PS services in UMTS. It provides both U-plane and C-plane PS-related functions. In the control plane, these functions include the GPRS mobility management (GMM) and session management (SM) functions. The SGSN also contains a subscriber database, similar to the VLR, which serves as intermediate storage for subscriber data to support subscriber mobility. It provides necessary control and signaling functionality to both UTRAN and GGSN and it may optionally communicate with the MSC with specific interfaces. Like GGSN, it also adheres to the OAM functionality, and, unlike SGSN, it collects the charging information related to radio network usage by the user. The procedure for information transfer between the SGSN, the GGSN, the VLR, and the HLR is defined in [32] and [33]. The 2G-SGSN and 3G-SGSN that are used in GSM/GPRS and UMTS systems, respectively, have some specific functions in addition to common ones. For example, the 2G-SGSN provides mobility management to the level of a cell, whereas 3G-SGSN provides the same to the level of an RNC for connected mode mobiles. On the other hand, they both detunnel GTP packets from the GGSN.

4.2.1.2   R99 Interfaces

The key interfaces between the MS and RAN, radio access and core networks, and between the CN elements of R99 are shown in Table 4.4. The details for each of these interfaces can be found in GSM/GPRS and R99 standards.

4.2.1.3   Call Processing and Intersystem Handoffs

The call flow for a R99 basic speech conversation is shown in Figure 4.16. It shows the functions of CCCH and DCCH in the process and also indicates the RAB assignments. The process of establishing the packet data services in UMTS is very similar to that of GPRS—through the activation of a PDP context, QoS, and so forth, as discussed in Chapter 2 [34].

The WCDMA systems have the capability to hand off to GSM/GPRS networks when UMTS coverage is not available. Performing handoffs of voice and data calls from the 5-MHz WCDMA RF channel to a 200-kHz TDMA system has unique challenges. These include interfrequency handoffs since GSM/GPRS and WCDMA operate on different frequencies, additional interference, and a reduction in capacity and coverage due to increased power usage. There are two possible approaches: the dual receiver approach and the slotted downlink transmission approach.

The handoffs are usually based on the slotted-mode (compressed-mode) technique, which allows the MS to adapt its transmission rate to allow for brief pauses in data transmission. These brief pauses allow the MS to monitor the transmission and transfer to GSM/GPRS radio channels. These small time periods are called transmission gap lengths (TGLs). TGLs are created by lowering the data rate (e.g., changing the speech coder rate) and thus reducing the spreading factor or puncturing the error correction code [3, 4].

Table 4.4 GSM/GPRS and R99 Interfaces

Interfaces	Elements
A	BSS and MSC. The signaling protocol on this interface is BSSAP.
B	MSC and VLR
C	HLR and GMSC. Used to exchange routing information, location information, and so forth.
D	MSC and HLR
E	MSC and MSC (MSCs of same or different infrastructure suppliers).
F	MSC and EIR. Used to verify the status of the IMEI received from MS.
G	VLR and VLR
Gb	BSS and SGSN
Gc	HLR and GGSN
Gf	SGSN and EIR
Gi	GGSN and PDNs
Gn	SGSN and GGSN
Gp	SGSN and GGSN
Gr	SGSN and HLR
Gs	MSC and SGSN
H	HLR and AuC. Used by the HLR when it needs to receive authentication data for a particular MS.
Iu	Core Network and UTRAN
Iub	Node-B and RNC
Iu-cs	UTRAN and MSC
Iu-ps	UTRAN and SGSN
Iur	RNC and RNC
Um	MS and BSS
Uu	MS and UTRAN

4.2.2 3GPP Release 4

As stated earlier, Rel-4 was the first 3GPP release to introduce technologies such as EDGE and TD-SCDMA. These two and other key Rel-4 features are briefly described in this section.

4.2.2.1 Rel-4 Features

Support for GSM/EDGE Radio Access Network
Besides the native GSM BSS and UTRAN radio access networks, Rel-4 also supports the GSM/EDGE radio access network (GERAN). GERAN is the radio part of EDGE technology. The EDGE concept, a new TDMA-based technology for IS-136 and GSM systems, provides 3G capabilities [35]. In Rel-4 GERAN is connected to the core network using an evolved version of the A and Gb interfaces. An overall description of GERAN can be found in [36].

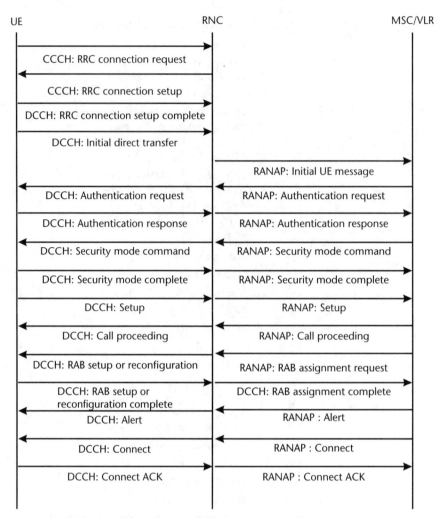

Figure 4.16   UMST basic call flow diagram [34].

EDGE is capable of offering data rates up to 384 Kbps, which is primarily done for 8-phase shift keying (8-PSK). EDGE does not require a new spectrum, and it is also known as enhanced GPRS (EGPRS) because it can coexist with GPRS, provided the necessary upgrades are implemented. It uses the same network elements, interfaces, protocols, and procedures as GPRS with minor differences, primarily in the RAN and MM.

EDGE introduces new coding schemes that are termed Modulation and Coding Scheme-1 to Modulation and Coding Scheme-9 (MCS-1 to MCS-9), as shown in Table 4.5. EGPRS also permits resegmenting (retransmitting with another coding scheme), which is not possible with GPRS. The resegmenting allows packets sent with higher coding that are received incorrectly to be dispatched again with lower coding schemes. EDGE uses 8-PSK for the upper five of its nine modulation and coding schemes, whereas GMSK is used for rest of the schemes. The channel types applicable to EGPRS are the same as those applicable to GPRS and thus the channels are shared among GPRS and EDGE users. To migrate from GPRS to EGPRS, carriers

Table 4.5   Modulation and Coding Schemes for EGPRS [34]

Scheme	Modulation	Input Data Payload (bits)	Data Rate (Kbps)
MCS-1	GMSK	176	8.8
MCS-2	GMSK	224	11.2
MCS-3	GMSK	296	14.8
MCS-4	GMSK	352	17.6
MCS-5	8-PSK	448	22.4
MCS-6	8-PSK	592	29.6
MCS-7	8-PSK	2*448	44.8
MCS-8	8-PSK	2*544	54.4
MCS-9	8-PSK	2*592	59.2

only need to upgrade the base stations; the core network does not require upgrading. In addition, new mobiles will be required to decode/encode the new modulation and coding schemes to carry the higher user data rates [34].

Protocol Stack.   The transmission plane protocol structure for GPRS is shown in Figure 4.17, which is also used by EGPRS with minor protocol revisions [37, 38]. The physical layer is split into a physical link sublayer (PLL) and a physical RF sublayer (RFL). The PLL provides functions such as data unit framing, data coding, and the detection and correction of physical medium transmission. The RFL performs modulation and demodulation of the physical waveforms. The RLC/MAC layer provides services for information transfer over the physical layer of the GPRS. It defines the procedures that enable multiple terminals to share a common transmission medium, which may consist of several physical channels.

In the network, the LLC is split between the BSS and SGSN. The BSS functionality is called LLC relay. Between the BSS and SGSN, the BSSGP conveys routing and QoS-related information and operates above the frame relay. Between GGSN and

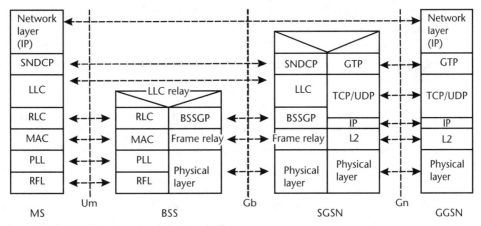

Figure 4.17   GPRS protocol architecture [37].

the SGSN, the GTP tunnels the packet data units through the GPRS backbone network by adding routing information. Below the GTP, the Transmission Control Protocol/User Datagram Protocol (TCP/UDP) and the Internet Protocol (IP) are used as the GPRS backbone network layer protocols. Ethernet-based or asynchronous transfer mode (ATM)–based protocols may be used below IP depending on the operator's network architecture. The Subnetwork Dependent Convergence Protocol (SNDCP) that runs between SGSN and MS maps network-level protocol characteristics onto the underlying LLC and provides functionalities such as multiplexing of network layers.

Support for TD-SCDMA

TD-SCDMA or, as it is also called, the low chiprate TDD option, is China's home-grown technology. The TD-SCDMA radio interface is integrated in 3GPP Rel-4 as the 1.28-Mcps option of the WCDMA-TDD, also called TDD low chiprate (TDD_{LCR}). TDD_{LCR} uses the UMTS core network and the TD-SCDMA air interface (see Chapter 5 for details). The details of the TD-SCDMA network architecture are described in Chapter 5.

Support for a Variety of Mobile Equipment

The Rel-4 network architecture supports a variety of handheld devices ranging from voice-only (circuit-switched) handsets to multimedia terminals and data aircards (for laptops).

Improved Speech Support in the CS Domain

Rel-4 features enhanced speech support, for example, with transcoder-free operation and tandem-free operation.[4] These enhancements provide transmission efficiencies and cost reductions in the core network and are applicable to both GSM and UMTS systems.

4.2.2.2   Rel-4 Network Architecture

Figure 4.18 shows the reference architecture of 3GPP Rel-4. As mentioned earlier, Rel-4 supports two new radio technologies, GERAN and TD-SCDMA, in addition to GSM BSS and UTRAN. To support GERAN and TD-SCDMA[5] services, few new components and interfaces have been added that do not exist in R99. Another important difference from R99 is that the MSC is split into two functional elements, the MSC server and media gateway function (MGW). The details of these core network elements are described next.

- *MSC server:* The MSC server is a signaling element that provides the call control and mobility control of the MSC and it operates in the C-plane. The MSC server sets up and manages the mobile originated and mobile terminated calls

---

4.   Tandem free operation refers to a configuration of a speech or multimedia call for which transcoders are not utilized but are present
5.   Some additional specific details of TD-SCDMA can be found in Chapter 5.

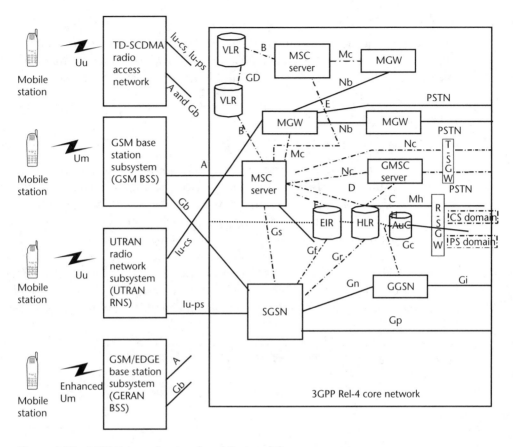

Figure 4.18 3GPP Release 4 network architecture [2].

over the CS domain. It also contains the VLR, which maintains the subscription data.

- *Media gateway function:* The MGW resides in the U-plane and handles the internetworking with the PSTN. It terminates bearer channels from an external circuit-switched network and media streams from a packet-switched network. It interacts with the MSC server and gateway MSC (GMSC) server for resource control by means of the H.248 protocol. The MGW includes the necessary resources for supporting UMTS/GSM transport media.

- *Gateway MSC server:* This element comprises the call control and mobility control parts of a GMSC.

- *Signaling gateway (SGW):* The SGW performs signaling translation at the transport level between Signaling System 7–based[6] transport signaling and IP-based transport signaling. Note that it does not perform any signaling translations at the application layer. It has two components, the transmitting SGW (T-SGW) and receiving SGW (R-SGW). The T-SGW provides signaling translation between the MSC server and PSTN, whereas the R-SGW interacts with the HLR for R99 or pre-R99 subscription data.

---

6. SS7 is the protocol designed for the transfer of control information between network elements.

4.2.2.3   Rel-4 Interfaces

Release 4 also introduces some new interfaces to provide interactions between the new and legacy network elements. Table 4.6 shows the interfaces that are introduced in Rel-4; their details can be found in respective standards.

4.2.3   3GPP Release 5

4.2.3.1   Rel-5 Features

Rel-5 can be considered the first release to introduce a new concept called the IP multimedia subsystem (IMS) or IP-based services. In addition to basic IP connectivity, session initiation capabilities are provided by the new IMS. Open service access (OSA) is being enhanced,[7] thus allowing subscribers to choose between service providers for enhanced IP multimedia services beyond basic services by opening the network to third-party application providers in a nonchaotic way. OSA is a middleware framework that abstracts the capabilities of the underlying network (e.g., providing user location information). Third parties can then access these capabilities in a controlled and secure way without requiring a detailed understanding of how the capabilities are provided. This breakthrough will allow many innovative multimedia services to be provided. The other key feature of Rel-5 is HSDPA, which provides very high data rates in the downlink. These two and other key Rel-4 features are briefly described here [2, 39]:

Table 4.6   Rel-4 Interfaces

Interfaces	Interacting Elements
A	GERAN and MGW
Gb	GERAN and SGSN
Iu-cs	TD-SCDMA and MSC
Iu-ps	TD-SCDMA and SGSN
Mc	MSC server and MGW and GMSC server and MGW. Supports different call models and provides an open architecture.
Mh	HLR and R-SGW. Supports the exchange of mobility management and subscription data information between the HLR and R99 or pre-R99 networks.
Nb	MGW and MGW. Used to support the U-plane traffic (speech traffic and so on) between MGWs.
Nc	MSC server and GMSC server and MSC server and T-SGW. Across the Nc interface, network-to-network call procedures are performed.
Um	MS and GERAN
Uu	MS and TD-SCDMA

7.   OSA was initially introduced in R99.

- *Support for IMS:* Rel-5 supports a special core network subsystem, which is called the IMS. The IMS is virtually a signaling system on top of the PS domain that enables the provision of IP multimedia services. It provides the network elements and procedures to support real-time and multimedia IP applications. It uses the Session Initiation Protocol (SIP) to support signaling and session control for real-time services.

- *HSDPA:* In Rel-5, techniques such as adaptive modulation and coding and H-ARQ can facilitate HSDPA. The details can be found in [40] and [41].

- *GERAN Iu Interface:* Rel-5 introduces the Iu interface for GERAN, which allows GERAN to connect directly to the UMTS core network.

- *IP transport in UTRAN:* Rel-5 enables the usage of IP technology as an alternative to ATM technology for the transport of signaling and user data over Iu, Iur, and Iub in the UTRAN. IP transport provides a cost-effective delivery alternative to its predecessor, ATM. Further information on this topic can be found in [42].

- *End-to-end QoS for the PS domain:* Rel-5 defines QoS [43], which provides a framework for end-to-end QoS. From the 3GPP perspective, the focus of end-to-end QoS is on the UMTS bearer (data) service only.

4.2.3.2 Rel-5 Network Architecture

Figure 4.19 shows the reference architecture of 3GPP Rel-5, which is very similar to Rel-4 architecture, except for IMS. All of the new major components are a so-called part of IMS. At the heart of IMS is SIP, which provides a centric signaling

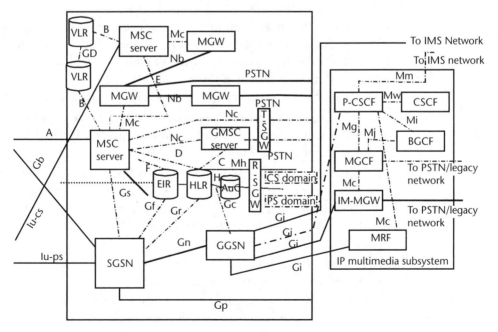

3GPP Rel-5 core network

Figure 4.19 Release 5 network architecture [2].

mechanism for the delivery of multimedia services over IMS. Thus, it becomes important to understand SIP before we look into the different components of IMS.

SIP has emerged as the key signaling and session control standard for the PS networks. It is compatible with both IPv4 and IPv6, which makes it an ideal choice for packetized voice and multimedia services. SIP was standardized by the Internet Engineering Task Force (IETF) in early 1999, and it has been chosen by both 3GPP and 3GPP2 as the signaling protocol for their packet telephony services. It is an application layer protocol that initiates, modifies, and terminates multimedia sessions over the Internet [44].

4.2.3.3 3GPP IMS

IMS is primarily developed for the mobile 3G and next generation devices communicating over IP with embedded SIP clients. It utilizes SIP signaling and control mechanisms with IMS-specific QoS and billing processes to offer a number of PS services. IMS supports VoIP as a basic service and a number of other multimedia applications. The basic idea behind IMS is to further enhance the Internet services currently provided by cellular technologies and provide a single platform instead of existing multiple ones [39].

The service requirements of the 3GPP IMS are specified in [45] and the technical details can be found in [46]. The IMS provides all of the network entities and procedures to support real-time voice and multimedia IP applications. It uses SIP to support signaling and session control for real-time services. The main functional entity in an IMS is the call state control function (CSCF) as shown in Figure 4.19.

- *CSCF*: CSCF manages the IP multimedia sessions by using the SIP protocol and acts as a SIP server. Three roles defined for a CSCF element:
  - *Proxy CSCF (P-CSCF)*: Communicates directly with the mobile terminals (SIP user agents) through the Gi interface, and it is the first point of contact in the IMS. It receives the requests issued by mobiles, and possibly after translation, it forwards them to another CSCF.
  - *Serving CSCF (S-CSCF)*: Provides the session control services. An IMS network might have more than one S-CSCF element depending on the number of subscribers and the capabilities required. It could also behave as a user agent and terminate SIP sessions.
  - *Interrogating CSCF (I-CSCF)*: Hides the S-CSCF addresses and internal topology of the service provider network from external entities. It queries the home subscriber services (HSS) to identify S-CSCF requirements, selects which S-CSCF form that the user receives services, and invokes the S-CSCF.
- *Media gateway control function (MGCF)*: Responsible for signaling between the PS domain and CS networks (e.g., PSTN).
- *Breakout gateway control function (BGCF)*: Selects to which PSTN a session should be forwarded. It then forwards the signaling to the appropriate MGCF and BGCF in the destination PSTN network.

- *IP multimedia media gateway (IM-MGW):* Performs media translation, that is, translation between media signals encoded in one format on one side and signals encoded in another format on the other side.
- *Multimedia resource function (MRF):* Is split into the MRF controller (MRFC) and MRF processor (MRFP). The MRFC controls the media stream resources of the MRFP, which processes and mixes the actual media streams.
- *HSS:* Is a database that stores subscription and location information. It can be thought of as an enhanced version of the home location register found in GSM.

4.2.3.4   Rel-5 Interfaces

Table 4.7 shows the interfaces that were introduced in Rel-5 and basically with IMS. These interfaces can be grouped into the following four main groups. The details of these interfaces can be found in the Rel-5 specifications.

4.2.4   Release 6

The key additions of Release 6 are as follows [47]:

- Support for HSUPA to improve the data rates and throughputs for uplink;
- Support for receiving diversity for HSDPA terminals;

Table 4.7   Rel-5 Interfaces

Interfaces	Interacting Elements
SIP-Based Signaling and Service Control Interfaces	
Mg	CSCF and MGCF
Mi	CSCF and BGCF. Carries SIP signaling and allows an S-CSCF to forward a multimedia session to BGCF and coordinate with PSTN.
Mj	BGCF and MGCF. Allows a BGCF to forward a multimedia session coming from external PSTN network to a MGCF.
Mk	BGCF and BGCF. Allows a BGCF to forward a session signaling to another BGCF.
Mr	S-CSCF and MRFC
Mw	P-CSCF and I-CSCF/S-CSCF. Is the primary signaling interface in IMS and is based on SIP.
Interfaces for Controlling Media Gateways	
Mc	MGCF and IM-MGW. Allows a signaling gateway (MGCF or MSC Server) to control a media gateway (IM-MGW or MGW).
Mp	MRFC and MRFP
Interfaces with the Information Servers	
Cx	CSCF and HSS
Interfaces with External Networks	
Go	P-CSCF and GGSN
Mb	MRFP and external IP network
Mm	CSCF and external IP network

- Support for multimedia broadcast/multicast service;
- Enhancement of beam-forming and remote control of antenna tilting;
- Introduction of IMS phase 2, which includes interworking between the IMS and CS and non-IMS IP networks, IMS group management (messaging and conferencing), and so forth;
- Improvements for UMTS/WLAN interworking;
- Enhancements to support end-to-end QoS;
- Security enhancements, including network domain security, subscriber certificates, and generic authentication architecture;
- GERAN improvements, including additional frequency bands to GSM, flexible layer 1, and uplink TDOA location determination;
- Improvements in OAM&P.

4.2.5   Releases 7 and 8

As stated earlier, Rel-7 and Rel-8 are still in the works. These are expected to provide services such as 3G Long-Term Evolution, MBMS, network sharing, priority service, WLAN/UMTS interworking, and enhancements to IMS [23].

4.3   High-Speed Downlink Packet Access

HSDPA has evolved from the WCDMA technology and is based on techniques such as adaptive modulation and coding and H-ARQ, which are also part of 1xEV-DO and 1xEV-DV.[8] These techniques are used to achieve high throughput, reduce delay, and provide high peak data rates to support a variety of applications. HSDPA relies on a new type of transport channel, the high-speed downlink shared channel (HS-DSCH), which is terminated in Node-B. HS-DSCH is applicable only to the PS domain's RABs. HSDPA is expected to achieve peak data rates of 14.4 Mbps in the downlink [48–52].

4.3.1   Basic Structure of HS-DSCH

4.3.1.1   Protocol Structure

Most of the protocols of the WCDMA are retained in HSDPA. The PDCP, RLC, and MAC-d layers are unchanged from the R99 and Rel-4 architecture. The new functionalities of H-ARQ and HS-DSCH scheduling are included in the MAC sublayer. In the UTRAN, these functions are included in a new entity called MAC-hs located in Node B. Two MAC protocol configurations are possible on the UTRAN side:

- *Configuration with MAC-c/sh:* In this configuration (Figure 4.20), the MAC-hs in Node-B (BTS) is located below MAC-c/sh in the CRNC.

---

8.   1x refers to a single carrier, EV-DO stands for evolution data optimized, and EV-DV refers to evolution data and voice.

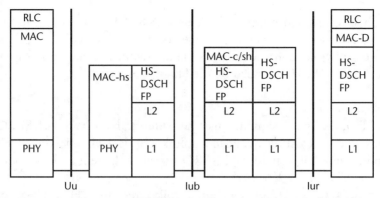

Figure 4.20 Protocol architecture of HS-DSCH, configuration with MAC-c/sh [48].

MAC-c/sh provides functions to HS-DSCH identical to those provided for the DSCH inR99. The HS-DSCH Frame Protocol (FP) will handle the data transport from the SRNC to the CRNC (if the Iur interface is involved) and between the CRNC and Node-B (Iub).

- *Configuration without MAC-c/sh:* In this configuration (Figure 4.21), the CRNC does not have any U-plane functions for the HS-DSCH. MAC-d in the SRNC is located directly above MAC-hs in Node-B: that is, in the HS-DSCH U-plane, the SRNC is directly connected to Node-B, thus bypassing the CRNC.

The same architecture supports both FDD and TDD modes of operation, though some details of the associated signaling for HS-DSCH are different.

4.3.1.2 Basic Physical Structure

Downlink
HS-DSCH applies to both the FDD and TDD modes of the physical layer. In the FDD mode, the basic downlink channel configuration consists of one or several high-speed physical downlink shared channels (HS-PDSCHs) along with the associated DPCH. These HS-PDSCHs are combined with a number of separate

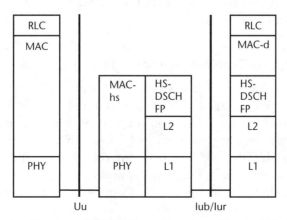

Figure 4.21 Protocol architecture of HS-DSCH, configuration without MAC-c/sh [48].

high-speed shared control channels (HS-SCCHs). In the TDD mode, the overall downlink signaling structure is based on associated dedicated physical channels and shared physical control channels. The downlink signaling information for support of the HS-DSCH is carried by the HS-SCCH.

Uplink

In FDD, the uplink signaling uses an additional DPCCH with SF = 256 that is multi-plexed with existing dedicated uplink physical channels. The HS-DSCH related uplink signaling consists of H-ARQ acknowledgment and channel quality indicator (CQI).

In TDD, the UE uses a shared uplink resource (HS-SICH) for transmitting an acknowledgment/negative ACK (ACK/NACK) and CQI information. The relation-ship between the HS-SCCH in the downlink and HS-SICH in the uplink is prede-fined and is not signaled dynamically on the HS-SCCH.

4.3.1.3 HS-DSCH Characteristics

Some of the characteristics of the HS-DSCH transport channels are as follows:

- HS-DSCH is processed and decoded from one coded composite traffic channel (CCTrCH). There is only one CCTrCH per HS-DSCH type, and each CCTrCH can be mapped to one or several physical channels.
- They are always associated with a DPCH and one or more HS-SCCHs.
- They exist only in the downlink and can possibly support beam-forming and link adaptation techniques and also broadcast in the entire cell.
- They support QPSK and 16-QAM modulation schemes.
- They support the turbo coding rate of 1/3.
- They assign a transmission time interval (TTI) of 2 ms for FDD. The TDD has two TTIs: For 3.84 Mcps, it is 10 ms, and for 1.28 Mcps, it is 5 ms. The TTI is also referred as a *subframe*.

4.3.2 MAC Architecture

This section describes the architecture of MAC required to support HS-DSCH on the UE and UTRAN sides.

4.3.2.1 HS-DSCH Architecture on the UE Side

Figure 4.22 shows the UE-side MAC architecture with HS-DSCH. It shows an addi-tional MAC entity called MAC-hs to support HS-DSCH functionality. The MAC-hs is configured via the MAC control SAP by RRC, similar to the MAC-c/sh and MAC-d, to set parameters in the MAC-hs, such as allowed transport format combi-nations for the HS-DSCH. It is also important to note that the associated downlink signaling carries information for support of HS-DSCH, whereas the associated uplink signaling carries feedback information.

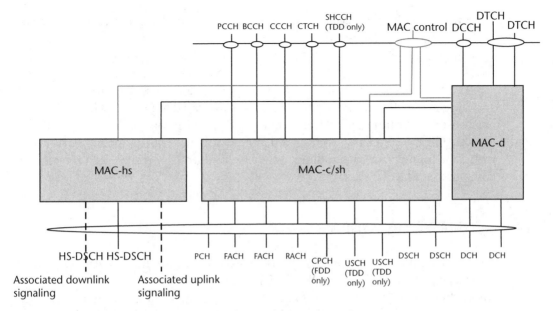

Figure 4.22 UE-side MAC architecture with HS-DSCH [48].

The MAC-d entity is modified with the addition of a link to the MAC-hs entity, whereas the MAC-c/sh is not modified for HS-DSCH. The MAC-hs handles the HS-DSCH specific functions. The MAC-hs comprises H-ARQ and reordering functions.

4.3.2.2 HS-DSCH Architecture on the UTRAN Side

Figure 4.23 shows the UTRAN-side MAC architecture. If an HS-DSCH is assigned to the UE, the MAC-hs SDUs (i.e. MAC-d PDUs) can be transmitted in two ways:

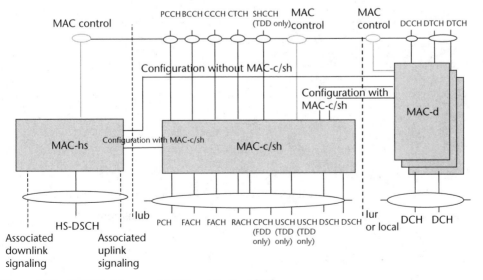

Figure 4.23 UTRAN-side overall MAC architecture [48].

• If transferred from MAC-c/sh to the MAC-hs via the Iub interface when configured with MAC-c/sh;

• If transferred from MAC-d via Iur/Iub when configured without MAC-c/sh.

The MAC-hs is responsible for handling the data transmitted on the HS-DSCH and the management of the physical resources allocated to HS-DSCH. The MAC-hs receives configuration parameters from the RRC layer via the MAC control SAP. The MAC-hs is comprised of four different functional entities including flow control, scheduling/priority handling, H-ARQ, and transport format and resource combination (TRFC) selection.

4.4 WCDMA Versus HSDPA

Table 4.8 highlights some of the key attributes of WCDMA and HSDPA. One significant change in HSDPA from WCDMA (R99) is that the TTI is reduced to 2 ms (three slots) from 10, 20, 40, or 80 ms. This reduction in frame size will shorten the round-trip delay between the UE and Node-B and it will also increase the efficiency of the adaptive modulation and coding scheme.

4.5 Summary

In summary, this chapter provided details of the WCDMA access technology. It illustrated the corresponding radio interface including the layers and the protocols. The chapter also provided information on the evolution of WCDMA, that is, HSDPA and the features of Releases 4 and 5. Readers are encouraged to go through the UMTS standard specifications and stay tuned to the HSPA (HSDPA plus HSUPA) and 3G-LTE industry announcements.

Table 4.8 WCDMA Versus HSDPA

Attributes	WCDMA-FDD	HSDPA
Total spectrum per carrier	10 MHz (5 MHz in each uplink and downlink)	10 MHz (5 MHz in each uplink and downlink)
Downlink peak data rate	2 Mbps	14.4 Mbps
TTI	10, 20, 40, or 80 ms	2 ms
Number of slots/TTI	15	3
Modulation	QPSK	QPSK and 16-QAM
Channel coding	Convolutional and turbo coding	Convolutional and turbo coding
Adaptive modulation and coding	No	Yes
Fast power control	Yes	No

References

[1] "Radio Interface Protocol Architecture," Technical Specification (Release 1999), 3GPP TS 25.301 (V3.11.0), Technical Specification Group Radio Access Network, 3GPP, September 2002.

[2] Willie, W. L., *Broadband Wireless Mobile*, New York: John Wiley & Sons, 2002.

[3] Garg, V. K., *Wireless Network Evolution*, Upper Saddle River, NJ: Prentice Hall, 2002.

[4] Lawrence, H., R. Levine, and R. Kikta, *3G Wireless Demystified*, New York: McGraw-Hill, 2002.

[5] "Spreading Modulation (FDD)," Technical Specification (Release 1999), 3GPP TS 25.213 (V3.9.0), Technical Specification Group Radio Access Network, 3GPP, December 2003.

[6] "Multiplexing and Channel Coding (FDD)," Technical Specification (Release 1999), 3GPP TS 25.212 (V3.4.0), Technical Specification Group Radio Access Network, 3GPP, September 2000.

[7] "Physical Layer Procedures (FDD)," Technical Specification (Release 1999), 3GPP TS 25.214 (V3.12.0), Technical Specification Group Radio Access Network, 3GPP, March 2003.

[8] "Services Provided by the Physical Layer (FDD)," Technical Specification (Release 1999), 3GPP TS 25.302 (V3.16.0), Technical Specification Group Radio Access Network, 3GPP, September 2003.

[9] "Spreading and Modulation (TDD)," Technical Specification (Release 1999), 3GPP TS 25.223 (V3.8.0), Technical Specification Group Radio Access Network, 3GPP, March 2002.

[10] "Multiplexing and Channel Coding (TDD)," Technical Specification (Release 1999), 3GPP TS 25.222 (V3.10.0), Technical Specification Group Radio Access Network, 3GPP, September 2002.

[11] "Physical Layer Procedures (TDD)," Technical Specification (Release 1999), 3GPP TS 25.214 (V3.13.0), Technical Specification Group Radio Access Network, 3GPP, September 2003.

[12] Lozhkin, A. N., "WCDMA Operation in TDD and FDD Modes: Evaluation of the Difference in BER Performances," *Proc. 2002 World Wireless Congress*, San Francisco, CA, May 28–31, 2002, pp. 219–224.

[13] Ahmed, I., "Scrambling Code Generation for WCDMA on the StarCore SC140/SC1400 Cores," *Freescale Semiconductor*, Rev. 1, 2004.

[14] Ojanpera, T., and R. Prasad, *Wideband CDMA for Third Generation Mobile Communications*, Norwood, MA: Artech House, 1998.

[15] Dahlman, E., et al., "WCDMA—The Radio Interface for Future Mobile Multimedia Communications," *IEEE Trans. on Vehicular Technology*, Vol. 47, No. 4, 1998, pp. 1105–1118.

[16] Wu, H., et al., "WCDMA-TDD for Multimedia Mobile Communications," *Proc. 2000 IEEE Int. Conf. on Communication Technology*, Beijing, China, August 21–25, 2000, Vol. 1, pp. 737–740.

[17] "Medium Access Control (MAC) Protocol Specification," Technical Specification (Release 1999), 3GPP TS 25.321 (V3.17.0), Technical Specification Group Radio Access Network, 3GPP, June 2004.

[18] "Radio Link Control (RLC) Protocol Specification," Technical Specification (Release 1999), 3GPP TS 25.322 (V3.18.0), Technical Specification Group Radio Access Network, 3GPP, June 2004.

[19] "Packet Data Convergence Protocol (PDCP) Specification," Technical Specification (Release 1999), 3GPP TS 25.323 (V3.10.0), Technical Specification Group Radio Access Network, 3GPP, September 2002.

[20] "Broadcast Multicast (BMC)," Technical Specification (Release 1999), 3GPP TS 25.324 (V3.9.0), Technical Specification Group Radio Access Network, 3GPP, December 2004.

[21] 3GPP TS 25.331 (V3.21.0), "Radio Resource Control (RRC) Protocol Specification," Technical Specification (Release 1999), Technical Specification Group Radio Access Network, 3GPP, December 2004.

[22] Xu, X., et al., "Mobile Network Evolution: A Revolution on the Move," *IEEE Communications Magazine*, Vol. 40, No. 4, April 2002, pp. 104–111.

[23] "Evolution of 3GPP Systems," Technical Report (Release 6), 3GPP TR 21.902 (V6.0.0), Technical Specification Group Services and Systems Aspects, 3GPP, September 2003.

[24] "Signaling Transport Mechanism Specification for the Base Station System—Mobile-Services Switching Centre (BSS-MSC) Interface," Technical Specification (Release 1999), 3GPP TS 08.06 (V8.0.1), Technical Specification Group GSM/EDGE Radio Access Network, 3GPP, May 2002.

[25] "Mobile-Services Switching Centre—Base Station System (MSC-BSS) Interface—Layer 3," Technical Specification (Release 1999), 3GPP TS 08.08 (V8.0.1), Technical Specification Group GSM/EDGE Radio Access Network, 3GPP, September 2003.

[26] "General Packet Radio Service (GPRS); Base Station System (BSS)—Serving GPRS Support Node (SGSN) Interface—Network Service," Technical Specification (Release 1999), 3GPP TS 08.16 (V8.0.1), Technical Specification Group GSM/EDGE Radio Access Network, 3GPP, May 2002.

[27] "General Packet Radio Service (GPRS); Base Station System (BSS)—Serving GPRS Support Node (SGSN); BSS GPRS Protocol (BSSGP)," Technical Specification (Release 1999), 3GPP TS 08.18 (V8.12.0), Technical Specification Group GSM/EDGE Radio Access Network, 3GPP, May 2004.

[28] Bettstetter, H., Vogel, H., and Eberspacher, J., "GSM Phase 2+ General Packet Radio Service GPRS: Architecture, Protocols, and Air Interface," *IEEE Communications Surveys*, Vol. 2, No. 3, 1999.

[29] "Network Architecture," Technical Specification (Release 1999), 3GPP TS 23.002 (V3.6.0), Technical Specification Group Services and Systems Aspects, 3GPP, September 2002.

[30] "UICC-Terminal Interface; Physical, Electrical and Logical Test Specification," Technical Specification (Release 1999), 3GPP TS 31.120 (V3.1.0), Technical Specification Group Terminals, 3GPP, December 2004.

[31] Cai, J., and J. D. Goodman, "General Packet Radio Service in GSM," *IEEE Communications Magazine*, Vol. 35, No. 10, October 1997, pp. 122–131.

[32] "Subscriber Data Management; Stage 2," Technical Specification (Release 1999), 3GPP TS 23.016 (V3.10.0), Technical Specification Group Core Network, 3GPP, March 2004.

[33] "General Packet Radio Service (GPRS); Service Description; Stage 2," Technical Specification (Release 1999), 3GPP TR 23.060 (V3.16.0), Technical Specification Group Services and Systems Aspects, 3GPP, December 2003.

[34] Smith, C., and J. Meyer, *3G Wireless with WiMax and Wi-Fi: 802.16 and 802.11*, New York: McGraw-Hill, 2004.

[35] Mazur, S., et al., "EDGE: Enhanced Data Rates for GSM and TDMA/136 Evolution," *IEEE Personal Communications*, Vol. 6, No. 3, June 1999, pp. 56–66.

[36] "Overall Description—Stage 2," Technical Specification (Release 5), 3GPP TR 43.051 (V5.10.0), Technical Specification Group GSM/EDGE Radio Access Network, 3GPP, August 2003.

[37] "Digital Cellular Telecommunications System (Phase 2+); General Packet Radio Service (GPRS); Service Description; Stage 2," Technical Specification (Release 98), 3GPP TS 03.60 (V7.9.0), Technical Specification Group Services and Systems Aspects, 3GPP, September 2002.

[38] Mazur, S., et al., "System Overview and Performance Evaluation of GERAN—The GSM/EDGE Radio Access Network," *Wireless Communications and Networking Conference*, September 23–28, 2000, Chicago, IL., Vol. 2, pp. 902–906.

[39] Varma, V., and K. Wong, "Supporting Real-Time IP Multimedia Services in UMTS," *IEEE Communications Magazine*, Vol. 41, No. 3, November 2003, pp. 148–155.

[40] "UTRA High Speed Downlink Packet Access (HSDPA)," Technical Specification (Release 4), 3GPP TS 25.950 (V4.0.0), Technical Specification Group Radio Access Network 3GPP, March 2001.

[41] "High Speed Downlink Packet Access (HSDPA) lub/lur Protocol Aspects," Technical Specification (Release 5), 3GPP TS 25.877 (V5.1.0), Technical Specification Group Radio Access Network, 3GPP, Jun. 2002.

[42] "IP Transport in RAN," Technical Report (Release 5), 3GPP TR 25.933 (V5.4.0), Technical Specification Group Radio Access Network, 3GPP, December 2003.

[43] "End-to-End Quality of Service (QoS) Concept and Architecture," Technical Specification (Release 5), 3GPP TS 23.207 (V5.9.0), Technical Specification Group Services and Systems Aspects, 3GPP, March 2004.

[44] Schulzrinne, H., and J. Rosenberg, "The Session Initiation Protocol: Internet-Centric Signaling," *IEEE Communications Magazine*, Vol. 38, No. 10, October 2002, pp. 134–141.

[45] "Core Network Subsystem (Stage 1)," Technical Specification (Release 5), 3GPP TS 22.228 (V5.6.0), Technical Specification Group Services and Systems Aspects, 3GPP, June 2002.

[46] "IP Multimedia Subsystem (IMS); Stage 2," Technical Specification (Release 5), 3GPP TS 23.228 (V5.13.0), Technical Specification Group Services and Systems Aspects, 3GPP, December 2004.

[47] "Full List of Release 6 Features," Meeting #28, 3GPP TSGS#28(05)0407, Technical Specification Group Services and Systems Aspects, 3GPP, June 2005.

[48] "High Speed Downlink Packet Access (HSDPA) Overall Description," Technical Specification (Release 6), 3GPP TS 25.308 (V6.3.0), Technical Specification Group Radio Access Network, 3GPP, December 2004.

[49] "High Speed Downlink Packet Access (HSDPA) Physical Layer Aspects," Technical Specification (Release 5), 3GPP TS 25.858 (V5.0.0), Technical Specification Group Radio Access Network, 3GPP, March 2002.

[50] "High Speed Downlink Packet Access (HSDPA) Enhancements," Technical Specification (Release 6), 3GPP TS 25.899 (V6.1.0), Technical Specification Group Radio Access Network, 3GPP, September 2004.

[51] "UTRA High Speed Downlink Packet Access (HSDPA)," Technical Specification (Release 4), 3GPP TS 25.950 (V4.0.0), Technical Specification Group Radio Access Network 3GPP, March 2001.

[52] "High Speed Downlink Packet Access (HSDPA) lub/lur Protocol Aspects," Technical Specification (Release 5), 3GPP TS 25.877 (V5.1.0), Technical Specification Group Radio Access Network, 3GPP, June 2002.

CHAPTER 5

Introduction to TD-SCDMA

In recent years, China has become the country with the largest and fastest growing cellular subscriber population in the world. The Chinese government anticipated this increase in its technology-savvy population and started working toward the development of its own wireless communications standard. The standardization effort took place in collaboration with the China Academy of Telecommunications Technology (CATT). The result of these efforts was time-division–synchronous CDMA (TD-SCDMA), a 3G standard completed by China Communications Standards Association (CCSA) and endorsed by the Chinese government. In May 2000, TD-SCDMA was accepted by the ITU, and later, in March 2001, it was accepted by 3GPP as a 3G standard and included in Release 4.

TD-SCDMA technology is integrated in 3GPP as the 1.28-Mcps option in IMT-TC and is called TDD low chiprate (TDD$_{LCR}$). Another technology that is part of IMT-TC is WCMDA-TDD, and both technologies have similar capabilities. TDD$_{LCR}$ uses the UMTS core network and the TD-SCDMA air interface. Thus, TD-SCDMA offers operators another choice besides WCDMA to upgrade their existing GSM networks. This chapter focuses on the TD-SCDMA system, its radio aspects, and its network architecture.

5.1 TD-SCDMA System

TD-SCDMA systems will use the architecture shown in Figure 5.1, which is similar to WCDMA's overall architecture. It has three main entities: user equipment (UE), a radio access network (RAN), and a core network (CN). The interface between the UE and RAN is Uu and that between the RAN and CN is Iu. The figure also illustrates high-level functional grouping into the access stratum and nonaccess stratum. The access stratum offers services through three SAPs to the NAS: general control (GC), notification (Nt), and dedicated control (DC). The SAPs are marked in circles in the figure and described in Chapter 4. Next we look into the radio aspects of the technology.

5.2 Radio Interface Protocol Architecture

The radio interface is layered into three protocol layers:

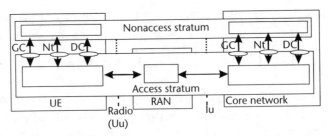

Figure 5.1   TD-SCDMA architecture [1].

- The physical layer (L1);
- The data link layer (L2);
- The network layer (L3).

Figure 5.2 shows the overall protocol architecture. Each layer is characterized by the services it provides and functions that support them. Layer 2 is split into two sublayers, MAC and RLC. L3 and RLC are divided into control and user planes. In the C-plane, L3 is partitioned into sublayers where the lowest sublayer, denoted as radio resource control (RRC), interfaces with layer 2. In the C-plane, the interface between RRC and higher L3 sublayers (mobility management and call control) is defined by GC, Nt, and DC SAPs. The signaling functions such as MM and CC belong to the NAS, and are outside the scope of this chapter.

The physical layer offers different transport channels to MAC. A transport channel is characterized by how information is transferred over the radio interface. MAC offers different logical channels to RLC. A logical channel is characterized by the type of information transferred. The physical channels are defined in the physical layer and are characterized by the code, frequency, and timeslot.

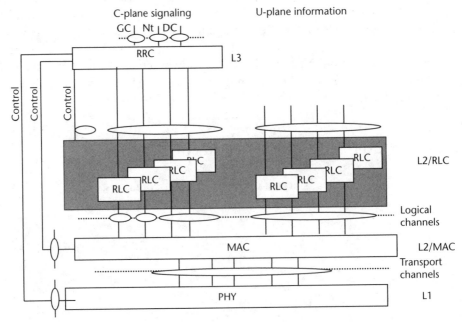

Figure 5.2   TD-SCDMA radio interface protocol architecture [2].

5.2.1   Physical Layer

The main function of the physical layer is to offer data transport services to higher layers. The access to these services is achieved through the use of transport channels via the MAC sublayer. TD-SCDMA uses a DS-CDMA technique (e.g., CDMA2000). Using DS-CDMA, the information is spread over approximately 1.6 MHz in TDD with a 200-kHz carrier raster for operating with unpaired bands. In the TDD, timeslots in a physical channel are divided into transmission and receiving parts, and the information on the forward link and reverse link is transmitted reciprocally.

There is also a TDMA component in the multiple access in addition to DS-CDMA. Thus, the multiple access has also been termed *TDMA/CDMA* due to the added TDMA nature. In addition, there is an option to use opportunity-driven multiple access (ODMA) for relaying information between nodes. Finally, most of the functions of L1 are similar to the L1 functions of WCDMA, which is discussed in Chapter 4.

5.2.1.1   Transport Channels

The physical layer offers data services to higher layers via transport channels. The transport channels are classified into two groups [1–3]:

1. *Common transport channels* where there is a need for in-band identification of the UEs when particular UEs are addressed;
2. *Dedicated transport channels* where the UEs are identified by the physical channel, that is, the code, timeslot, and frequency for TDD.

Common Transport Channels
The common transport channels are as follows:

- *Random access channel (RACH):* A contention-based uplink channel used for transmission of relatively small amounts of data, for example, for initial access, nonreal-time dedicated control, or traffic data;
- *Forward access channel (FACH):* A downlink channel with closed-loop power control used for transmission of relatively small amounts of data;
- *Downlink shared channel (DSCH):* Shared by several UEs carrying dedicated control or traffic data;
- *Uplink shared channel (USCH):* Shared by several UEs carrying dedicated control or traffic data;
- *Broadcast channel (BCH):* A downlink channel used for broadcast of system information into an entire cell;
- *Synchronization channel (SCH):* A downlink channel used for broadcast of synchronization information into an entire cell;
- *Paging channel (PCH):* A downlink channel used for broadcast of control information into an entire cell, allowing efficient UE sleep mode procedures;
- *ODMA random access channel (ORACH):* A contention-based channel used in the relay link.

Dedicated Transport Channels

The dedicated transport channels are as follows:

- *Dedicated channel (DCH):* Dedicated to only one UE, either in the uplink or downlink;
- *ODMA dedicated channel (ODCH):* Dedicated to only one UE in the relay link.

With each transport channel, there is an associated transport format (for transport channels with a fixed or slow changing rate) or an associated transport format set (for transport channels with a fast changing rate). A transport format is defined as a combination of encodings, interleaving, bit rate, and mapping onto physical channels. A transport format set is a set of transport formats, for instance, a variable-rate DCH has a transport format set for each rate, whereas a fixed-rate DCH has a single transport format.

5.2.1.2   Physical Channels

As stated in the previous sections, the physical channels are recognized by the code, frequency, and timeslot. The transport channels are mapped onto physical channels to carry the information on the physical medium. The physical channels are as follows:

- *Dedicated physical channel (DPCH):* Transfers user data on the uplink and downlink channels. The dedicated traffic channel (DTCH) is part of the DPCH.
- *Common control physical channel (CCPCH):* Two types, and both of these use a fixed SF of 16:
  - *Downlink common control physical channel (DCCPCH):* Mapped to the BCH, PCH, or FACH;
  - *Physical random access channel (PRACH):* Mapped to the RACH.
- *Physical synchronization channel (PSCH):* Located in both the downlink and uplink. The downlink PSCH is located in the DwPTS (downlink pilot), and the uplink PSCH is located in the UpPTS (uplink pilot).

Frame Structure

TD-SCDMA uses both TDMA and CDMA techniques to allocate and differentiate communication channels. All physical channels have a four-layer structure of superframes, radio frames, subframes, and timeslots/codes (Figure 5.3). Each timeslot consists of guard symbols, and the codes are used like a TDMA component to separate different user signals in the time and code domains.

The superframe has 72 frames with each frame having a duration of 10 ms, which is further divided into 2 subframes of 5 ms each (Figure 5.3). Each subframe consists of seven main timeslots (TS) of 0.675 ms and three special timeslots: DwPTS (downlink pilot), G (guard period), and UpPTS (uplink pilot). The frames are used to carry both user data and dedicated signaling. A particular timeslot is

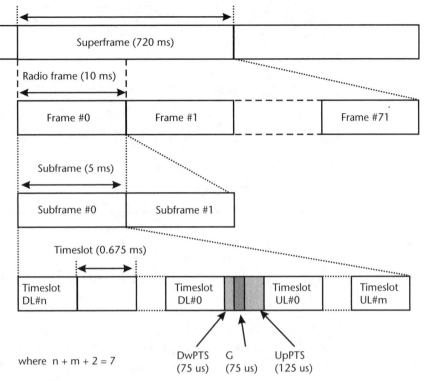

Figure 5.3   Physical channel structure [3].

called a *burst,* which is a physical channel in the TDD. A burst consists of two data symbol fields, a midamble of 144 chips and a guard period of 16 chips. During transmission, a burst is first encoded with a unique OVSF channelization code. The OVSP codes allow multiple bursts at the same time. The burst is then encoded with a common scrambling code. The superframe of TD-SCDMA exactly matches the WCDMA (both FDD and TDD solutions) multiframe, thus ensuring compatibility [3–6].

5.2.1.3   Mapping of Transport Channels to Physical Channels

Figure 5.4 shows the mapping between transport and physical channels.

5.2.1.4   Multiplexing and Channel Coding

A data stream from/to MAC and higher layers is encoded/decoded to create transport services over the radio transmission link. The channel coding scheme is a combination of error detection, error correction, rate matching, interleaving, and mapping transport channels onto/splitting from physical channels.

In the TD-SCDMA mode, the total number of basic physical channels per subframe provides the maximum number of timeslots and CDMA codes per timeslot. Figure 5.5 illustrates the overall transport channel coding and multiplexing. The data are received by the coding/multiplexing unit in the form of transport block sets once every transmission time interval. The time interval is variable (10, 20, 40,

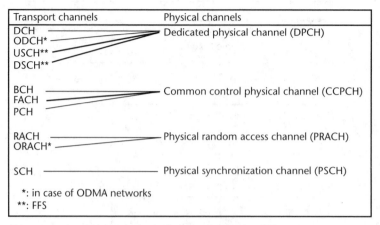

Figure 5.4   Transport channel—to—physical channel mapping [3].

or 80 ms) depending on the transport channel. The steps of the coding/multiplexing are described below [4, 7]:

- *Error detection:* The first step is error detection, which uses the commonly known cyclic redundancy check (CRC). The CRC is 24, 16, 8, or 0 bits in length, which is determined by higher layers for each transport channel.
- *Transport block concatenation and code block segmentation:* First the transport blocks are serially concatenated, and then code block segmentation is applied. The maximum size of the code block depends on the type of the coding scheme that is used for the transport channel.
- *Channel coding:* There are three channel coding schemes that can be applied to the code blocks: convolution, turbo, and none. The type of channel coding scheme is dependent on the transport channel. Table 5.1 shows coding schemes and coding rates that are used by different transport channels.

The remaining process consists of radio frame size equalization, first interleaving, radio frame segmentation, rate matching, transport channel multiplexing, physical channel segmentation, second interleaving, and finally physical channel mapping. These steps are briefly described as follows:

- Radio frame size equalization involves padding the input bit sequence to ensure that the output can be segmented in data segments of the same size.
- The first interleaving is a block interleaver with intercolumn permutations.
- Radio frame segmentation is applied when the TTI is longer than 10 ms. The input bit sequence is segmented and mapped onto consecutive radio frames.
- Rate matching means that bits on a transport channel are repeated or punctured. A rate matching attribute assigned by the higher layers is used when the number of bits to be repeated or punctured is known.
- During the transport channel multiplexing phase, one radio frame from each channel is sent for multiplexing every 10 ms. These radio frames are multiplexed into a coded composite transport channel (CCTrCH).

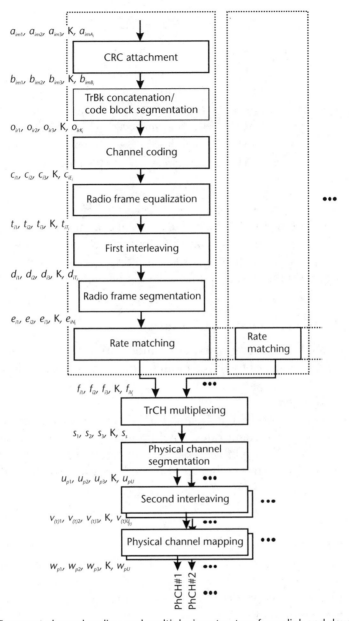

Figure 5.5   Transport channel coding and multiplexing structure for uplink and downlink [7].

- When more than one physical channel is used, physical channel segmentation divides the bits among the different physical channels.
- The second interleaving can be applied jointly to all data bits transmitted during one frame or separately within each timeslot on which the CCTrCH is mapped. The selection of the second interleaving scheme is controlled by a higher layer.
- The bits after physical channel mapping are denoted by w_{p1}, w_{p2}, ..., w_{pUp}, where p is the physical channel number and U_p is the number of bits in one

Table 5.1   Error Correction Coding Parameters [7]

Transport Channel	Coding Scheme	Coding Rate
BCH	Convolutional code	1/2
PCH	Convolutional code	1/2
FACH	Convolutional code	1/2
RACH	Convolutional code	1/2
DCH	Convolutional code	1/3, 1/2, or no coding
DCH	Turbo code	1/3 or no coding

radio frame for the respective physical channel. The bits w_{pk} are mapped to the physical channels so that the bits for each physical channel are transmitted over the air in ascending order with respect to k. The mapping scheme depends on the applied second interleaving scheme.

5.2.1.5   Modulation and Spreading

TD-SCDMA provides two types of modulation: data modulation and spreading modulation. Some of the basic modulation parameters are shown in Table 5.2.

Data Modulation
The data modulation is QPSK, and TD-SCDMA provides an option to use 8-PSK for 2 Mbps and higher data rates. The modulation is performed at the output of the physical channel mapping block of the transport channel multiplexing and channel coding structure. The symbol chiprate and duration are defined as follows:

$$T_s = Q \times T_c \qquad (5.1)$$

where $T_c = 1/\text{chiprate} = 0.78125$ µs, and the symbol time T_s depends on the spreading factor Q. Up to $K = 16$ CDMA codes can be assigned to a single user or multiple users who are simultaneously using the same timeslot or frequency. The number of CDMA codes is dependent on the spreading factor, the interference scenario, and the service requirements.

Pulse shape filtering is applied to each chip at the transmitter before modulation. A root-raised cosine is the impulse response of the pulse shape filter. This means that the overall function at the receiver should also be a root-raised cosine.

Table 5.2   Basic Modulation Parameters [8]

Chiprate	1.28 Mcps
Carrier spacing	1.6 MHz
Data modulation	QPSK or PSK (optional)
Chip modulation	Root-raised cosine roll-off $\alpha = 0.22$
Spreading characteristics	Orthogonal Q chip/symbol where $Q = 2^p$, $0 <= p <= 4$

Spreading Modulation.
Each data symbol will have a spreading code c of varying length (spreading factor) Q. The spreading factor varies from 1 to ~16; that is, $Q = \{1, 2, 4, 8, 16\}$. The data symbol is spread with OVSF codes, which are defined in Chapter 4. Each level in the OVSF code tree defines spreading factors. All codes in the code tree cannot be used simultaneously in a given timeslot.

Once the spreading code is obtained, a scrambling sequence is applied to a specific cell. Scramble codes are used to identify different cells. A set of 22 codes can be used as the scramble codes. A scramble code is cell specific; its index is the same as the synchronization (SYNC) code index, which means that scramble codes and SYNC codes have a one-to-one relationship.

5.2.1.6   Physical Layer Services

Newer technologies such as uplink synchronization, smart antennas, joint detection, baton handover, and power control have been adopted in the TD-SCDMA standard. This section provides a high level understanding of some of these new functions and services as applicable to TD-SCDMA [4, 5, 9, 10].

Transmitter Power Control
The transmitter power control (TPC) is applied to limit the interference level within the TD-SCDMA system. The TPC helps in reducing both the intracell and intercell interference level and to reduce the power consumption in the UE. Some of the parameters of TPC are shown in Table 5.3.

It is worth noting that all of the codes within one timeslot allocated to the same bearer service use the same transmission power. Also, if a user has simultaneous real-time and nonreal-time traffic, closed-loop power control is used for both services; however, depending on the services, different power levels are used.

Smart Antennas
TD-SCDMA base stations are equipped with smart antennas that target signals to and from specific terminals and reduce intercell interference. The base station tracks the mobiles throughout the cell, so that the signal-to-interference ratio of the mobile terminal is improved by several decibels. In addition, smart antennas optimize the link budget and increase the capacity of the TD-SCDMA systems.

Joint Detection/Terminal Synchronization
Joint detection eliminates the multiple access interference (MAI) typically associated with multiuser access, through parallel processing of individual traffic streams.

Table 5.3   TPC Characteristics [3]

Attribute	Uplink	Downlink
Dynamic range	80 dB	30 dB
Power control rate	Variable	Variable
	Closed loop: 0–20 cycles/s	Closed loop: 0–200 cycles/s
	Open loop: 1–7 slots delay	
Step size	[1...3] dB	[1...3] dB

Terminal synchronization improves the uplink signal's quality by accurately tuning the transmission timing of each individual terminal with respect to its base station. The efficient combination of joint detection and terminal synchronization enhances the coverage of a base station and enable operators to use TD-SCDMA deployments for macrocell, microcell, and picocell applications.

Random Access Procedure

When the UE is in the idle mode, it will maintain downlink synchronization and read the cell broadcast information. From the cell broadcast message, the UE will get the code set assigned to UpPTS SYNC1 for random access, the number and position of the RACH channel, the number and position of the FACH channel, the operation mode (symmetric or asymmetric) of the cell, and other information related to random access. Also, the UE needs to estimate the timing and power level for the transmission of random access bursts according to the received DwPTS from Node-B. Finally, the UE and Node-B exchange the information and packets related to access in the FACH/RACH pair mentioned earlier and finish the random access procedure in the physical layer.

Cell Search Procedure

After locating the cell, the UE determines the DwPTS, subframe, and superframe synchronization of that cell and then decodes the contents in BCH. The initial cell search is carried out in three steps:

Step 1: Search for DwPTS.   During this step, the UE uses the SYNC (in DwPTS) to acquire DwPTS synchronization in a cell. This task is accomplished using one or more filters that are matched to the received SYNC, which in turn is selected from the Gold sequences set. The DwPTS synchronization detects peaks in the matched filter(s) output(s). The corresponding SYNC of the maximum peak is the SYNC of the strongest cell.

Step 2: Subframe Synchronization.   During the second step, the UE receives the BCH, which is followed by the DwPTS. Once the SYNC is detected, the BCH scramble code can be determined, and the UE can verify the correction of the BCH. According to the result of verification, UE may go to the next step or go back to step 1.

Step 3: Superframe Synchronization.   During the final step, the UE searches for the head of the superframe and gets the frame number first. Then it reads the complete broadcast information of the found cell into one or several BCHs.

Baton Handover

Baton handover depends on the positioning ability of the MSs of the smart antenna system. It utilizes the advantages of both soft and hard handover procedures. Due to multipath problems in dense urban areas, a single BS is not good enough to determine the accurate position of a mobile. Thus the system relies on the reporting information of the mobile during the cell search procedure to determine the target BS. Both intrahandover and interhandover procedures are supported by Baton handover. The basic procedure of Baton handover is as follows:

1. During the first step, the UE obtains the relevant information from nearby cells by listening to the broadcast information of its current BS.

2. The UE selects a target BS using the broadcast information of its current BS.

3. The UE then sends the relevant information to the target BS through some common transport channel so that the target BS can locate the mobile.

4. Finally, either the UE or Node-B can then initiate the handover procedure.

5.2.2   Media Access Control Sublayer

This section highlights the architecture and functions of the MAC sublayer. The MAC architecture is constructed from the MAC entities, which perform functions that are different in the UE from those completed in the RAN. These entities are as follows [11]:

- *MAC-b:* Handles the BCH. There is one MAC-b entity in each UE and one in the RAN for each cell.

- *MAC-c:* Handles the FACH, the RACH, and the PCH. There is one MAC-c entity in each UE and one in the RAN for each cell.

- *MAC-d*: Responsible for handling of dedicated logical channels and dedicated transport channels allocated to a UE. There is one MAC-d entity in each UE and one in the RAN for each cell.

- *MAC-sh*: Handles DSCHs and USCHs. There is one MAC-sh entity in each UE that is using a DSCH and a USCH and one MAC-sh entity in the RAN for each cell that contains a DSCH and a USCH.

- *MAC-sy*: Handles the information received on the SCH.

5.2.2.1   MAC Entities

The MAC entities and their architecture related to the UE and RAN sides are described in this subsection.

MAC-b, MAC-c, and MAC-sy
Figure 5.6 shows the connectivity of the MAC-b, MAC-c, and MAC-sy entities in a UE and in each cell of the RAN. MAC-b, MAC-c, and MAC-sy represent BCH, PCH, and SCH control entities, respectively, which are cell-specific MAC entities in the RAN. In the UE side there is one SCH and BCH and one PCH control entity per UE. The SCH control entity handles synchronization channels. The MAC control SAP is used to transfer control information to each MAC entity.

Traffic-Related Architecture—UE Side
Figure 5.7 shows the connectivity of MAC-c, MAC-d, and MAC-sh entities. The figure shows that MAC-d maps several DTCHs to a number of DCHs; MAC-sh controls access to a common transport channel, and MAC-c is connected with the MAC-d for transfer of data and radio network temporary identities (RNTIs). The

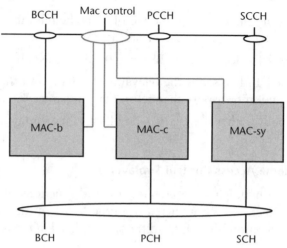

Figure 5.6 UE-side and RAN-side architecture [11].

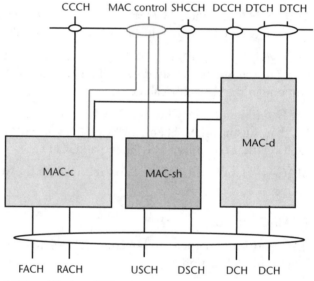

Figure 5.7 UE-side MAC architecture [11].

MAC control SAP is used to transfer control information to each MAC entity. The SHCCH SAP is used to transfer control information from RRC to MAC-sh. The MAC-sh transfers data from the DSCH to the MAC-d and from the MAC-d to the USCH under control of the FACH. Further details of these entities' functions can be found in [11].

Traffic-Related Architecture—RAN Side
Figure 5.8 describes the connectivity between the MAC entities from the RAN side. It is similar to the UE case with the exception that there will be one MAC-d for each UE and each UE (MAC-d) that is associated with a particular cell may be associated with that cell's MAC-sh. MAC-c and Mac-sh are located in the controlling RNC,

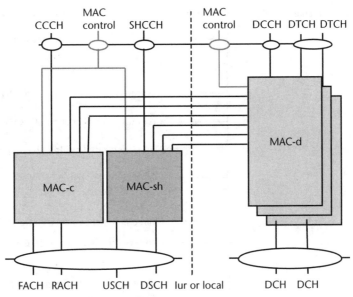

Figure 5.8   RAN-side MAC architecture [11].

whereas MAC-d is located in the serving RNC. The MAC control SAP is used to transfer control information to each MAC entity belonging to one UE. The SHCCH is used by the RRC in the controlling RNC to transfer control information to MAC-sh. Further details of these entities' functions can be found in [11].

5.2.2.2   Channel Structure

The MAC operates on both transport channels and logical channels. The transport channels (see Section 5.2.1.1) are described between MAC and layer 1, while logical channels are described between MAC and RLC.

Logical Channels
The MAC layer provides data transfer services on logical channels. A set of logical channel types is defined for the different kinds of data transfer services offered by MAC. As mentioned earlier, each logical channel type is defined by what type of information is transferred. There are two types of logical channels: control and traffic (Figure 5.9). The control channels are used for transfer of control plane information only, whereas the traffic channels are used for the transfer of user plane information only.

Mapping Between Logical and Transport Channels
The following mapping or connections exist between logical and transport channels:

- SCCH is connected to SCH.
- PCCH is connected to PCH.
- CCCH is connected to RACH and FACH.

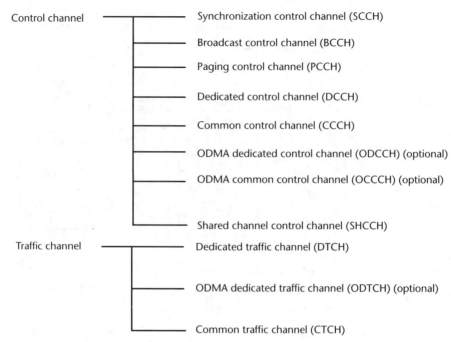

Figure 5.9 Logical channel structure [11].

- DCCH and DTCH can be connected to RACH and FACH, or to RACH and DSCH, or to DCH and DSCH, or to DCH.
- CTCH can be connected to FACH and DSCH or BCH.
- DCCH and DTCH can be connected to the USCH.

5.2.2.3 MAC Services and Functions

MAC provides data transfer services to the upper layers. The data transfer includes reallocation of radio resources and MAC parameters and reporting of measurements.

MAC functions are very similar to those provided by the MAC sublayer of WCDMA. The readers can find details in Chapter 4 and [12, 13].

5.2.3 Radio Link Control Sublayer

Figure 5.10 provides an overview model of the RLC sublayer. The figure shows different RLC peer entities for the three modes, namely, transparent mode (Tr), unacknowledged mode (UM), and acknowledged mode (AM). There is a separate transmitting and a receiving entity for the transparent mode service and the UM mode service and one combined transmitting and receiving entity for the AM service. The dashed lines between the AM entities illustrate the possibility of sending the RLC control data (e.g., resynchronization PDUs and acknowledgments) and data PDUs on separate logical channels. Data transfer through layer 2 is characterized by the three applied data modes on RLC in combination with the data transfer

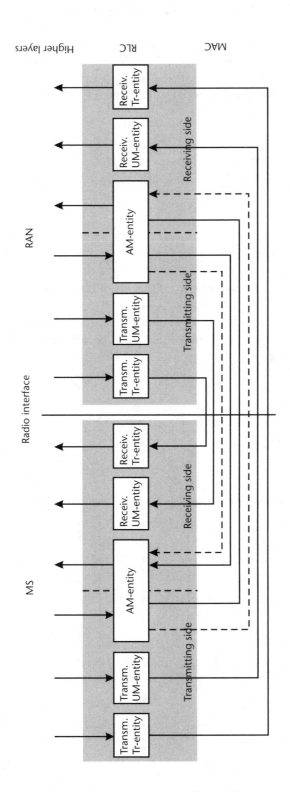

Figure 5.10 Overview of RLC [14].

type on the MAC, whether or not a MAC header is required. The transparent MAC transmission does not require a header, whereas both acknowledged and unacknowledged transmissions require a RLC header. In unacknowledged transmission, only one type of unacknowledged data PDU is exchanged between peer RLC entities. In acknowledged transmission, both (acknowledged) PDUs and control PDUs are exchanged between peer RLC entities. The details of the three entities are described in the next few sections.

5.2.3.1   RLC Entities

The RLC entities and their architecture are described in this section.

Transparent Mode Entities
As stated in the preceding section, the transparent mode has a pair of peer entities (Figure 5.11). The transmitting Tr entity receives SDUs from the higher layers through the Tr-SAP. The RLC may or may not segment the SDUs into appropriate PDUs; the decision is made when service is established. The RLC delivers the RLC PDUs to the MAC through logical channels as shown in Figure 5.11. The Tr entity receives PDUs through one of the logical channels from the MAC sublayer. The RLC reassembles (if segmentation has been performed) the PDUs into RLC SDUs; the reassembling process is decided on when the service is established. The RLC delivers the RLC SDUs to the higher layer through the Tr-SAP.

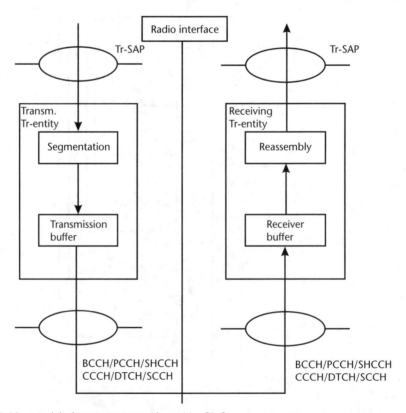

Figure 5.11   Model of transparent mode entities [14].

Unacknowledged Mode Entities
Similar to the Tr mode, the UM mode also has one transmitting and one receiving entity, as shown in Figure 5.12. The transmitting UM entity receives SDUs from the higher layers and either segments them into RLC PDUs or concatenates them with other SDUs, depending on their size. The RLC adds a header and the PDU is placed in the transmission buffer and delivered to the MAC sublayer using one of the logical channels, as shown in Figure 5.12. The receiving UM entity receives PDUs through one of the logical channels from the MAC sublayer. The RLC removes the header from the PDUs, reassembles the PDUs (if segmentation has been performed) into RLC SDUs, and delivers them to higher layer.

Acknowledged Mode Entity
Unlike the Tr and UM modes, the AM mode only has one entity, as illustrated in Figure 5.13. The transmission side of the AM entity receives SDUs from the higher layers and segments and/or concatenates them into payload units (PUs) of fixed length. PU length is a semistatic value that is decided in the bearer setup, and it can only be changed through bearer reconfiguration by RRC.

   After segmentation/concatenation, the PUs are placed in the retransmission buffer and transmission buffer. The RLC-PDU is constructed from PU buffers. The multiplexer (MUX) then selects PDUs and decides when these PDUs are delivered to MAC. For example, RLC control PDUs can be sent on one logical channel and data PDUs on another. The PDUs are delivered via a function that completes the

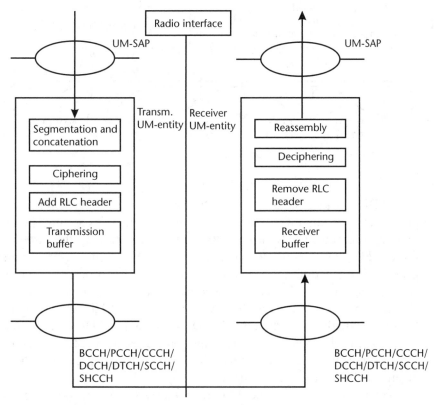

Figure 5.12   Model of unacknowledged mode entities [14].

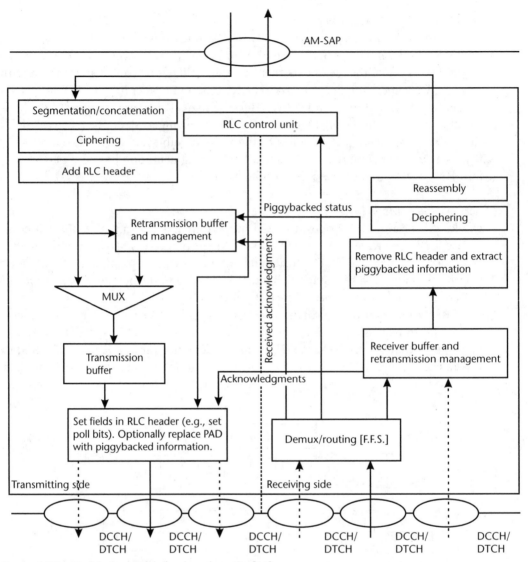

Figure 5.13 Model of acknowledged mode entity [14].

RLC-PDU header and potentially replaces padding (concatenation) with piggybacked status information. This function includes setting the poll bit, compressing subsequent PUs into one RLC-PDU, or setting up the extended RLC-PDU header (PUs not in sequence) where applicable. When the piggybacking mechanism is applied, the padding is replaced by control information that is not saved in any retransmission buffer. Instead it becomes part of the piggybacked STATUS PDU, which is in turn included in the AM data transfer PDU (AMD-PDU). The dashed lines illustrate the case where AMD-PDUs and control PDUs are transmitted on separate logical channels. The retransmission buffer also receives acknowledgments from the receiving side that are used to indicate retransmissions of PUs and when to delete a PU from the retransmission buffer.

The receiving side of the AM entity receives PDUs through one of the logical channels from the MAC sublayer. The RLC-PDUs are expanded into separate PUs and potential piggybacked status information is extracted. The PUs are placed in the receiver buffer until a complete SDU has been received. The receiver buffer requests retransmissions of PUs by sending negative acknowledgments to the peer entity. Subsequently, the headers are removed from the PDUs, the PDUs are reassembled into a SDU, and then the SDU is finally delivered to the higher layers. The receiving side also receives acknowledgments from the peer entity that are passed to the retransmission buffer on the transmitting side.

5.2.3.2   RLC Services to Upper Layers

This section provides an overview of the services provided by RLC to upper layers. The TD-SCDMA RLC sublayer provides all of those services that are provided by the WCDMA RLC sublayer. Therefore, readers are encouraged to refer to Chapter 4 and [1, 15, 16]. This section highlights only those services that are not part of the WCDMA RLC sublayer.

- *RLC connection establishment/release:* This service performs establishment/ release of RLC connections.
- *Unacknowledged data transfer—unique delivery:* The RLC sublayer delivers each SDU only once to the receiving upper layer, using a duplication detection function.

5.2.3.3   RLC Functions

This section highlights only those functions that are not part of the WCDMA RLC sublayer. Therefore, readers are encouraged to refer to Chapter 4 and [15, 16].

- *Connection control:* This function performs establishment, release, and main-tenance of a RLC connection.
- *Header compression:* This feature that includes several PUs into one RLC-PDU is referred to as RLC header compression. The RLC header can be applied for AM data transfer service, and its applicability is negotiable between the network and the UE. The application of RLC header compression is optional for the network, but it is mandatory for the UE.

5.2.4   Radio Resource Control Sublayer

The TD-SCDMA RRC sublayer entities, services, and functions are similar to those of the WCDMA RRC sublayer; therefore, they are not described in this chapter. The readers can find the details in Chapter 4 and [1, 12, 16, 17].

5.3   CCSA TD-SCDMA Network Architecture

TD-SCDMA is a new radio air interface that uses most of the core network elements of GSM or UMTS. If the TD-SCDMA radio network is connected to a GSM core, it is called TD-SCDMA for Mobile Communication (TSM); if it is hooked up to a UMTS core, it is called TDD_{LCR} [5, 6, 18].

- *TSM:* The TSM system utilizes the GSM core network, including its signaling and protocols. TSM is an enhancement of the existing GSM/GPRS core network with a TD-SCDMA radio subsystem, as shown in Figure 5.14. In this phase, the GSM protocol stack (layer 2–3) is used with the TD-SCDMA air interface. The new TD-SCDMA base stations (Node-Bs) are connected to the existing GSM BSC. The GSM BSC is upgraded to TD-SCDMA radio subsystems by means of the software TGCF. In this way the BSC migrates to a T-RNC as shown in Figure 5.15. The BSC (software upgraded to TD-SCDMA) is connected to the GSM core network by the existing A and Gb interfaces. This seamless integration of a 3G air interface into an existing and stable GSM infrastructure results in the short-term availability of 3G services without installation of a completely new core network infrastructure. Intersystem handover between TD-SCDMA and GSM/GPRS ensures seamless interworking between the two radio systems.
- *TDD_{LCR}:* TDD_{LCR} uses the UMTS core network and the TD-SCDMA air interface. The new TD-SCDMA base transceiver stations are connected to the TDD_{LCR} and, hence, enable the RNCs. These RNCs support the A and Gb interfaces to connect to MSC and GGSN/SGSN and the Iu (Iu-cs and Iu-ps) UMTS interfaces to connect to the UMTS core network (Figure 5.15). It also supports UMTS layer 2/layer 3 protocols.

5.3.1 Handover Information Transfer

The inter-radio access technology (TD-SCDMA to UMTS) handover information transfer procedure is used by the UE to convey the RRC information needed for hard handover to UTRAN (Figure 5.16). This procedure is activated when the UE picks up a UMTS signal that is better than the TD-SCDMA signal [19].

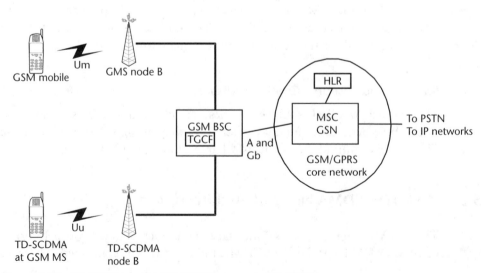

Figure 5.14   TD-SCDMA in the GSM/GPRS core network (TSM).

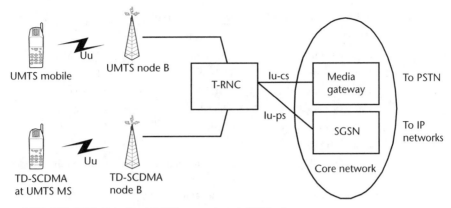

Figure 5.15   TD-SCDMA in the UMTS core network (TDD$_{LCR}$).

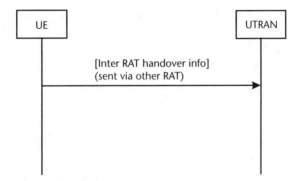

Figure 5.16   Inter-RAT handover information transfer [19].

5.3.2   Intersystem Handovers

The purpose of the intersystem handover procedure is to, under the control of the network, transfer a connection between the UE and UTRAN to TD-SCDMA. This procedure can be used in both circuit-switched and packet-switched domains, and the details can be found in [19].

In the case of TD-SCDMA/RAN to GSM/BSS transfer, the handover procedures for both circuit-switched and packet-switched domains are based on measurement reports from the UE but initiated from the RAN. The handover is considered successful when the UE is able to perform the intersystem handover from RAN connected mode to GSM connected mode. When the UE has sent a complete handover message to the GSM/BSS, it then initiates a temporary block flow toward GPRS and sends a radio access update request. A similar procedure is applied for GSM to TD-SCSMA handovers [20].

5.4   WCDMA-TDD Versus TD-SCDMA

Chapters 4 and 5 have shown that WCDMA-TDD and TD-SCDMA have similar capabilities. Table 5.4 summarizes some of the characteristics of these two technologies.

Table 5.4   WCDMA-TDD Versus TD-SCDMA

Attribute	WCDMA-TDD	TD-SCDMA
Access method	DS-CDMA	DS-CDMA
Duplex mode	TDD	TDD
Carrier bandwidth	5 MHz	1.6 MHz
Chiprate per carrier	3.84 Mcps	1.28 Mcps
Modulation	QPSK	QPSK, 8-PSK
Peak data rate	2 Mbps	2 Mbps
Channel coding	Convolutional code; turbo code	Convolutional code; turbo code
Frame structure	Multiframe: 720 ms; radio frame: 10 ms	Superframe: 720 ms; radio frame: 10 ms
Subframe	Not defined	5 ms
Timeslots per radio frame	15	14
Power control	Downlink and uplink: closed loop Uplink: open loop	Downlink and uplink: closed loop Uplink: open loop
Synchronization	Synchronous	Synchronous

5.5   Summary

This chapter presented an overview of the TD-SCDMA standard. It discussed the similarities between the UMTS and TD-SCDMA radio interfaces. It also provided a detailed description of the three layers of the TD-SCDMA radio interface and the integration processes of TD-SCDMA with GSM/GPRS and UMTS networks.

References

[1]   "Radio Interface Protocol Architecture," Technical Specification, Working Group 1, CWTS TS C001 (V3.0.0), Beijing: China Wireless Telecommunications Standard, October 1999.

[2]   "Physical Layer—General Description," Technical Specification, Working Group 1, CWTS TS C101 (V3.0.0), Beijing: China Wireless Telecommunications Standard, October 1999.

[3]   "Physical Channels and Mapping of Transport Channels onto Physical Channels," Technical Specification, Working Group 1, CWTS TS C102 (V3.0.0), Beijing: China Wireless Telecommunications Standard, October 1999.

[4]   Kammerlander, K., "Benefits and Implementation of TD-SCDMA," *Proc. IEE Int. Conf. on Communications*, Beijing, China, August 21–25, 2000, pp. 1013–1016.

[5]   Siemens, "TD-SCDMA—Going 3G and Beyond," 2001.

[6]   Siemens, "TD-SCDMA: the Solution for TDD bands," March 2002.

[7]   "Multiplexing and Channel Coding," Technical Specification, Working Group 1, CWTS TS C103 (V2.2.0), Beijing: China Wireless Telecommunications Standard, October 1999.

[8]   "Spreading and Modulation," Technical Specification, Working Group 1, CWTS TS C104 (V3.0.1), Beijing: China Wireless Telecommunications Standard, October 1999.

[9]   "Physical Layer Procedures," Technical Specification, Working Group 1, CWTS TS C105 (V3.0.0), Beijing: China Wireless Telecommunications Standard, October 1999.

[10] "Services Provided by the Physical Layer," Technical Specification, Working Group 1, CWTS TS C002 (V3.0.0), Beijing: China Wireless Telecommunications Standard, October 1999.

[11] "MAC Protocol Specification," Technical Specification, Working Group 1, CWTS TS C201 (V3.0.0), Beijing: China Wireless Telecommunications Standard, October 1999.

[12] "Radio Interface Protocol Architecture," Technical Specification (Release 1999), Technical Specification Group Radio Access Network, 3GPP TS 25.301 (V3.11.0), 3GPP, 2002.

[13] "Medium Access Control (MAC) Protocol Specification," Technical Specification (Release 1999), Technical Specification Group Radio Access Network, 3GPP TS 25.321 (V3.17.0), 3GPP, 2004.

[14] "RLC Protocol Specification," Technical Specification, Working Group 1, CWTS TS C202 (V2.1.0), Beijing: China Wireless Telecommunications Standard, October 1999.

[15] "Radio Link Control (RLC) Protocol Specification," Technical Specification (Release 1999), Technical Specification Group Radio Access Network, 3GPP TS 25.322 (V3.18.0), 3GPP, 2004.

[16] "RRC Protocol Specification," Technical Specification, Working Group 1, CWTS TS C203 (V2.1.0), Beijing: China Wireless Telecommunications Standard, October 1999.

[17] "Radio Resource Control (RRC) Protocol Specification," Technical Specification (Release 1999), Technical Specification Group Radio Access Network, 3GPP TS 25.331 (V3.1.0), 3GPP, 2004.

[18] Siemens, "Taking the Right Path Toward 3G," 2002.

[19] "Radio Resource Control (RRC) Protocol Specification," Technical Specification (Release 7), Technical Specification Group Radio Access Network, 3GPP TS 25.331 (V7.1.0), 3GPP, 2006.

[20] "UE Functions and Interlayer Procedures in Connected Mode," Technical Specification, Working Group 1, CWTS TS C003 (V3.0.0), Beijing: China Wireless Telecommunications Standard, October 1999.

CHAPTER 6

Wireless Local Loop

6.1 Introduction

Wireless local loop or WLL utilizes many similar, if not the same, concepts as those used in the cellular and PCS systems. It is, however, different than cellular or PCS systems in its application, which is *fixed* and not mobile. A WLL is sometimes called a *fixed cellular system*. WLL, also called radio in the loop (RITL) or fixed-radio access (FRA), is a system that uses a wireless connection for the last mile of delivering plain old telephone service (POTS) to customers.

WLL has been deployed in many developing countries to provide local telephone service where either the cost of traditional wireline (copper) local loop is prohibitive or the areas are so remote that it is not feasible to install wired services. In those parts of the world, WLL provides reliable, economical basic telephone service in place of traditional copper wireline. Furthermore, WLL technology solutions can be deployed much faster compared to wireline solutions, which commonly take months for deployment.

WLL can be operated in both licensed and unlicensed frequency bands, and so, unlike cellular, WLL does not have many spectrum constraints. Technology alternatives for WLL can be classified into four general categories: point-to-point systems, cellular systems modified for WLL applications, satellite-based systems, and systems designed specifically for WLL applications. The first two categories are elaborated on in this chapter.

Choosing the best WLL technology can be a challenging task, given the number of technologies available in the different frequency bands. Many WLL technologies are proprietary, that is, either limited to a specific frequency band or only available from a few manufacturers, making it difficult to achieve the economies of scale that will lead to subsequent equipment cost reductions. Further exaggerating the problem is the availability of wireless broadband services from cellular operators. Although there are not many cost-effective WLL options for delivering broadband services, some deployments have been made throughout the world to provide those services using 3G (WCDMA and CDMA2000) wireless technologies.

In this chapter we first look into basic WLL concepts and then review the three well-known international TDMA-based standards for low-mobility, low-power wireless communications applications, namely, Digital Enhanced Cordless Telecommunications (DECT), Personal Access Communication System (PACS), and the Personal Handyphone System (PHS). Next, we compare the capacity of TDMA- and CDMA-based WLL systems, and finally we will look into 3G technologies for

WLL and provide a WLL market update. This approach lays the groundwork for the fixed wireless broadband services (point-to-point systems and systems designed specifically for fixed wireless applications) that are described in the next chapter [1–3].

6.2   WLL Overview

In recent years, a number of manufacturers have developed customized WLL systems. In addition, the ITU has listed and elaborated on the parameters of a number of cellular mobile telecommunication standards that can be implemented in WLL Systems [3, 4]. These include most of the 2G and 3G technologies' standards. The earlier WLL systems used TDMA access technology for the most part, but with the advent of superior CDMA technology, the momentum has shifted toward CDMA-based 2G and 3G technologies (see Figure 6.1 and Figure 6.2).

6.3   Initial WLL Technologies Overview

DECT, PACS, and PHS are three well-recognized standards originally developed to provide mobile and portable radio telephone service to high-traffic microcellular and picocellular environments. However, the three standards are being implemented in WLL applications as a replacement for wireline access. These three low-tier systems are good for a small coverage area and provide better voice quality than the cellular systems. The three main reasons for their implementation in WLL applications are as follows [4]:

1. *Speech quality:* DECT, PACS, and PHS are based on 32-Kbps adaptive differential pulse code modulation (ADPCM) waveform coding. Speech quality is superior to other lower bit rate, speech-coded cellular radio standards, and it is comparable to wireline pulse code modulation (PCM).

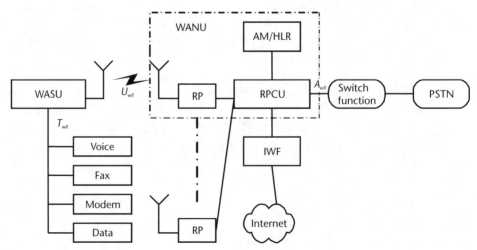

Figure 6.1   WLL system architecture.

Figure 6.2 Generalized network architecture for wireless-to-wireline data internetworking.

2. *ISDN connectivity:* These standards provide provisions for suitable connection to the ISDN.
3. *Capacity:* High capacity is the key factor in successful implementation of WLL systems in dense urban areas. DECT, PACS, and PHS are microcellular systems capable of providing much higher capacities than normal macrocellular systems and standards.

Table 6.1 provides a high-level view of these three standards [4, 7]. One common attribute among them is that they use TDMA as the access method, which makes them less attractive compared to CDMA as we will describe in the next section. Studies have shown that the three technologies are suitable for WLL service, but the most suitable one will depend on the specific environment and the given set of conditions.

6.4 TDMA Versus CDMA Access for WLL Application

This section compares the capacities of TDMA and CDMA access technologies. The 2G cellular systems, except for IS-95 and low-tier systems, use TDMA, whereas CDMA has become the major standard for the 3G mobile systems. A number of studies and field deployments have shown that CDMA provides better capacity than the TDMA systems.

The major difference in the capacity calculation of WLL and cellular systems is the value of the E_b/N_0 ratio. The E_b/N_0 ratio is the ratio of energy per bit to noise density; it can also be considered the signal-to-interference ratio. Since the environments of a mobile system and a WLL system are different, the value of the E_b/N_0 ratio is different, even if the systems are using the same access technology [2, 4].

Table 6.1 DECT, PACS, and PHS Attributes

Attribute	DECT	PACS	PHS
Frequency band (MHz)	1,880–1,900	1,850–1,910 1,930–1,990	1,895–1,918
Duplex method	TDD	FDD	TDD
i0Access method	TDMA	TDMA	TDMA
Number of carriers	10	200	77
Carrier separation (kHz)	1,728	300	300
Peak data rate (Kbps)	1,152	384	384
Modulation	GFSK	pi/4 QPSK	pi/4 QPSK
Speech coding	ADPCM	ADPCM	ADPCM
Cell size	Small	Large	Largest

Thus, note that capacity for a fixed WLL CDMA link is higher than that for a mobile CDMA because of the better power control in the former.

The capacity (M_t) of a TDMA WLL system is:

$$M_t = \left(B_w / b_c \right) / N \tag{6.1}$$

where:

B_w = total spectrum bandwidth;

b_c = channel bandwidth;

N = frequency reuse factor:

$$N = 1/3 \left[6 \left(E_b / N_o \right) \right]^{2/\gamma} \tag{6.2}$$

where E_b/N_0 is as defined earlier and γ is the propagation path-slope, which is assumed to be 4. The capacity (M_c) of a CDMA WLL system is:

$$M_c = \left\{ \left(G_p \right) / \left[3/\lambda \times E_b / N_o \times v_f \times \left(1 + \beta \right) \right\} \right] + 1 \tag{6.3}$$

where:

G_p = processing gain = bandwidth/data rate;

λ = sectorization gain (~2.55);

v_f = voice activity factor (~0.4);

β = adjacent cell interference (~0.6).

If we look at Table 6.2, it is very clear that WLL systems based on CDMA will provide better capacity than those that use TDMA technology. Also, because all of the 3G mobile standards are based on CDMA, the use of 3G technologies for WLL application is on the rise, as we will show in the later sections.

Table 6.2 TDMA Versus CDMA WLL Systems

Attribute	IS-95 (CDMA)	IS-136 (TDMA)	GSM (TDMA)
Channel bandwidth (kHz)	1,250	30	200
Number of carriers*	3	5,000/30 = 167	5000/200 = 25
E_b/N_0	6	14	12
Frequency reuse factor	1	7	3
Traffic channels per sector	3	= 167/(3 × 4) = 13.92	= 25/(3 × 3) = 2.78
Capacity per sector (# of users)	33	15	24
Capacity (Erlangs) per sector at 2% blocking	24.6	9.0	16.6

Note: Total bandwidth is assumed to be 5 MHz.

6.5   3G Technologies for WLL

The invention of the Internet and the need for high-speed data services have led to the deployment of 3G technologies. The need to access these features is also on the rise from the WLL customers. In the last few years, we have seen several deployments of WLL systems that are based on the WCDMA and CDMA2000 (1X and 1xEVDO) air interfaces. This section briefly describes the WCDMA- and CDMA2000-based WLL systems.

6.5.1   WCDMA for WLL

One of the 3G air interfaces that can be used for WLL is WCDMA. The air interface is the standard published by the TTA of Korea. This standard, called "Wideband CDMA Air Interface Compatibility Interim Standard for 2.3 GHz Band WLL System," was first published in October 1996 and updated in December 1997. This standard provides a basis for the implementation of the WLL radio system equipment, including the customer premises, base station, and controller systems. It applies to the wireless connection segment between the terminal and the base station. It specifies a WCDMA-based protocol operating in the 2.30- to 2.40-GHz band. The standard supports high-speed data service for telephone, facsimile, and ISDN through the WLL system for PSTN users. In particular, it specifies a wireless packet transmission mode to support these services efficiently through WLL to address the increasing demands of data and Internet users.

The standard allows two types of bandwidth (5 and 10 MHz). The standard includes basic requirements for physical layer (layer 1), data link layer (layer 2), and network layer (layer 3) of WLL wireless connection [5, 8].

The overall architecture of this WCDMA-based WLL system is very similar to that shown in Figure 6.1. The WCDMA radio interface defined in this standard is between the terminal and the base station. The terminal consists of the existing fixed PSTN, ISDN terminals, and the terminal interface module that connects the existing fixed terminals with the base station through the radio interface. The base station provides the function of network radio interface termination and the function of wired interface with the switching system. The switching system provides local switching or toll switching functions, and it is also connected to the fixed communication network, such as PSTN, ISDN. The interworking function (IWF) unit of the WLL system provides a gateway to external data networks, including the Internet, X.25, and FR networks. We provide some key aspects of the layering structure of this standard; note that details can be found in [9].

- *Physical layer aspects:* The WCDMA supports a physical layer for modulation/demodulation to provide telephone, facsimile, high-speed data, and packet communication service for existing PSTN users.
- *Data link layer aspects:* The DLL is used to deliver the data of the network layer with high reliability. Its main functions include connectionless mode operation, data resending, point-to-point data transfer between originating and terminating terminals, error control, and sequence control.

• *Network layer aspects:* The network layer performs the control procedure from power-on of the terminal until the assignment of a subscriber-dedicated radio channel resource. It also supports bearer mode change procedures between telephone and facsimile. It also performs the signaling function for bearer connection control and the signaling function for subscriber registration, authentication, encryption, attachment, and detachment.

6.5.2   CDMA2000 for WLL

Unlike WCDMA-based WLL, no particular standard defines CDMA2000 as the air interface for WLL. However, it is the beauty of the existing 3G CDMA2000 family of standards that it can be used for WLL without any significant modifications [3].

Many of the shortcomings experienced with traditional WLL networks can be avoided through the use of CDMA2000 technology. In wireless systems, cell coverage is governed by the laws of physics, more specifically by the propagation of the RF signal. PHS and DECT systems located in the 1,800/1,900-MHz band use spectrum bands similar to those of PCS systems. But their relatively low-power base stations support a smaller coverage area. The PHS standard specifies a maximum base station power output of up to 500 mW, which is equivalent to the microcell range of a regular mobile system. The typical radius of coverage for a PHS/DECT base station is on the order of 500m, and the maximum range is in the order of 1,500m. As a result, the large number of cells and the associated backhaul required for coverage, especially in rural areas, lead to substantial capital and operating costs. Some fixed wireless systems deployed in the 3.4- to 3.6-GHz band can provide cell coverage up to several kilometers by utilizing directive receive antennas on high masts. However, one big drawback of these systems is the line-of-sight (LOS) requirement, which ultimately impacts the number of base stations required and the associated capital and operating costs.

In contrast to these technologies, CDMA2000-based WLL systems support a variety of base stations, depending on the coverage need. Macrocell, microcell, and picocell sites, which can be deployed in any market (rural, suburban, urban, and dense urban areas), can provide coverage ranging from 10m to 100 km. The LOS requirement is not an issue, given the frequencies at which CDMA2000 operates. In the end, operating in non-LOS conditions and also at lower frequencies leads to the deployment of fewer cell sites. This results in faster deployments and lower capital and operating costs.

In terms of handsets or fixed wireless terminals, the core components that are used in CDMA2000 mobiles and WLL handsets are common to both, leading to considerable economies of scale. There are many CDMA-based fixed wireless terminal (FWT) manufacturers in the market today. A feature-rich CDMA2000 1x WLL phone usually includes a USB/RS232 port for Internet and data capabilities, messaging, and the standard landline phone features.

6.6   WLL Technologies Comparison

The terrestrial WLL technologies can be broadly divided into two broad categories: low-tier PCS or microcellular-based systems and cellular-based systems. Table 6.3

summarizes the capabilities of these options and the applications for which these are suited.

6.7 WLL Market Update

WLL has gained considerable acceptance in the last decade, especially in developing countries where the cost of basic landline telephony service is very cost prohibitive. Service providers across the globe have deployed both standard-based and customized (proprietary) WLL solutions. Many manufacturers have developed low-cost WLL systems and FWTs.

CDMA-based WLL systems have been deployed in about 40 countries around the global, mostly in developing countries [3, 10]. According to the CDMA Development Group, most of the CDMA (CDMAOne, CDMA2000) low-cost WLL systems that have been tested and deployed in Africa, Asia, and Latin America have been developed by Chinese manufacturers. The cost of WLL terminals has dropped, encouraging CDMA-based WLL deployments in developing regions. In the coming years, we could see more new CDMA2000-based WLL network deployments in Asia, Africa, Latin America, and Eastern Europe.

According to some studies, the overall WLL equipment market was expected to exceed $3 billion by 2006. The growth is expected due to deregulation and improving economic circumstances in the developing nations. If the trend continues, the WLL systems are likely to start replacing the landlines, as we are currently witnessing in the developed nations. Beyond this technology substitution, we will also start seeing the use of fixed wireless for broadband services in the developing nations, which leads us to the next chapter.

6.8 Summary

To summarize, an overview of WLL, its architecture, and TDMA-based WLL technologies (e.g., DECT, PACS, and PHS) was presented. A comparison of TDMA versus CDMA WLL capacities was also given, and it demonstrated that the WLL systems based on CDMA technology provide better capacity. We also described

Table 6.3 WLL Technologies Comparison

Features	Low-Tier System	Cellular-Based System
Systems	DECT, PACS, PHS	CDMAOne, CDMA2000 1X, WCDMA, 1xEV-DO
Application	Facilitate rapid market entry and expand capacity of existing infrastructure; provide data.	Expand basic telephony service; provide data.
Power	Low	High
Subscriber density	High	Medium and low
Coverage area	Small	Large

WCDMA- and CDMA2000-based WLL systems and provided market statistics for the WLL systems.

References

[1]   Smith, C., and J. Meyer, *3G Wireless with WiMax and Wi-Fi: 802.16 and 802.11*, New York: McGraw-Hill, 2004.

[2]   Garg, V., and E. Sneed, "Digital Wireless Local Loop System," *IEEE Communications Magazine*, Vol. 34, No. 10, October 1996, pp. 112–115.

[3]   Naidu, M., and V. Kripalani, "CDMA2000 for Wireless in Local Loop Networks," *Qualcomm*, December 2004, pp. 1–21.

[4]   Momtahan, O., and H. Hashemi, "A Comparative Evaluation of DECT, PACS, and PHS Standards for Wireless Local Loop Applications," *IEEE Communications Magazine*, Vol. 39, No. 5, May 2001, pp. 156–163.

[5]   Ulema, M., and Y. Yoon, "A Wideband CDMA based Wireless Local Loop Protocol," *Wireless Communications and Networking Conference, IEEE*, September 21–24, 1999, pp. 52–524.

[6]   Noerpel, A., and Y. Lin, "Wireless Local Loop: Architecture, Technologies and Services," *IEEE Personal Communications*, Vol. 5, No. 3, March 1997, pp. 74–80.

[7]   Ulema, M., et al., "Low-Tier Wireless focal Loop Radio Systems—Part 7: Introduction," *IEEE Communications Magazine*, Vol. 35, No. 3, March 1997, pp. 84–92.

[8]   Ulema, M., and Y. Yoon, "A Wireless Local Loop System Based on Wideband CDMA Technology," *IEEE Communications Magazine*, Vol. 37, No. 10, October 1999, pp. 128–135.

[9]   "Wideband CDMA Air Interface Compatibility for 2.3GHz Band WLL System (Layer1)," TTAS.KO-06.0015, Korea: TTA, December 1997.

[10] CDMA Development Group; http://www.cdg.org.

CHAPTER 7

Broadband Wireless Access

7.1 Introduction

One of focus areas in telecommunications is the delivery of efficient capacity at acceptable cost for the last mile of a connection. The last mile is the final point at which connectivity is delivered from a service provider to a customer. We will look into several technologies in this chapter that potentially provide solutions for last mile connectivity.

Chapter 6 illustrated the concepts of WLL while focusing on narrowband technologies such as fixed cellular systems for last mile access. In this chapter, we discuss the radio access technologies that will provide us with true fixed/portable and perhaps mobile (*in the next few years*) broadband data service. The broadband wireless access (BWA) systems will not only provide wireless high-speed data, but they can also be used as alternative wireless backhaul solutions for existing cellular systems. Cellular operators are currently adding more and more T1 lines to support the growth of their data traffic, which is as yet still in its infancy. As this data traffic grows, cellular operators will be looking for cost-effective but reliable alternatives to support this ongoing growth, and this is where BWA systems will come into play.

BWA systems are considered members of WLL or fixed wireless access (FWA) systems. BWA is emerging as a technology that features several techno-economic advantages over wireline solutions (for example, DSL and cable modems). On the other hand, BWA systems will not only require solid economics but, most importantly, mobility to drag some businesses away from the cellular (EV-DO and HSDPA) operators. Some BWA technologies operate at ultrahigh frequencies that require LOS conditions for their operation; others do not require LOS. In recent years, the IEEE has endorsed both LOS and non-LOS (NLOS) systems. Besides the IEEE-based systems, some proprietary BWA NLOS systems are also currently available.

In this chapter, we will describe both LOS and NLOS systems along with other BWA technologies that can operate in both licensed and unlicensed frequency bands. These technologies or systems are Local Multipoint Distribution Service (LMDS), the IEEE 802.16 Broadband Radio Systems (BRS), IEEE 802.20, ETSI Broadband Radio Access Network (BRAN), free space optics (FSO), and millimeter wave (E-band). Some key aspects of BRS are frequency spectrum, network topology, standards, and challenges; these topics are discussed in this chapter.

7.2   Local Multipoint Distribution Service

The LMDS is a fixed wireless service typically used in the 20- to 40-GHz range, depending on the country of use. LMDS is a wireless system that is used for the deployment of integrated bidirectional voice, video, and data services using the allotted frequency spectrum. At these frequencies, the range is limited to about 5 miles, hence the use of the term *local* in LMDS. Service providers are focusing on these systems to provide an alternative access solution to end users (last mile service) compared to copper, fiber, or cable. LMDS can be a very cost-effective alternative to a competitive local exchange carrier (CLEC). As with LMDS, a wireless service provider can deploy a wireless system without laying cables or copper in the ground to reach the customers. However, radio equipment is still very expensive at such high frequencies. The cost of LMDS equipment has to has to come down before it can become an attractive solution.

LMDS employs cellular features, such as design and the frequency reuse factor, with two key exceptions: There is no handoff, and it requires LOS conditions for operation. Also, LMDS can be considered a form of WLL (see Chapter 6). The difference is that signals at such frequencies suffer much greater attenuation due to rain, trees, and foliage, so LMDS systems require highly reliable modems [1].

LMDS is the name given by the U.S. researchers to terrestrial point-to-multipoint (PMP) broadband communications in the Ka-band (~30 GHz). In Canada, these systems are referred as Local Multipoint Communications Systems (LMCS). In Europe, these systems are categorized as Multipoint Video Distribution Services (MVDS), which is less appropriate because these systems deliver audio and data in addition to video. However, in last few years the term LMDS has begun to become more common in Europe.

7.2.1   LMDS Architecture

It is important to note both TDMA and FDMA multiplexing schemes can be used in the uplink. TDMA is a better choice for a large customer base that generates bursty traffic. FDMA is best suited for large companies where a large number of users can share one LMDS link.

7.2.2   LMDS Frequency Spectrum

There are unique differences in the service offerings allowed in Europe compared to the United States and rest of the world. Some of the major differences lie in the different requirements imposed in different countries and perhaps on specific operators. Depending on the country in which the system is deployed, the license will impose restrictions such as a minimum coverage requirement for a set of services over a specified period of time. Another difference is the spectrum assignments and the channel plans put forth by a specific country or the country's standards body.

In 1998, the U.S. Federal Communications Commission (FCC) allocated the 1.3-GHz bandwidth for LMDS (27.5 to 28.35 GHz, 29.1 to 29.25 GHz, and 31 to 31.3 GHz), as shown in Figure 7.1 for PMP services. In Europe, multiple frequency bands are available for LMDS services. The most common one operates between

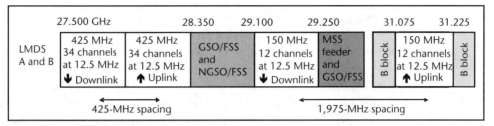

Figure 7.1 LMDS bands in the United States.

40.5 and 42.5 GHz. These systems also operate in the 26-GHz band (24.5 to 26.5 GHz and 27.5 to 29.5 GHz), and the 38-GHz band is also being deployed, which allows 37- to 39.5-GHz frequencies for such systems in Europe. Besides the United States and Europe, other countries have also set aside spectrum for LMDS in recent years; for example, Korea and Japan use frequencies from 22 to 28 GHz.

7.2.3 LMDS Standards

7.2.3.1 ETSI Standard

The ETSI completed one of the first LMDS specifications back in 1998. The title of these specifications was "Digital Video Broadcasting (DVB); Local Multipoint Distribution System (LMDS)." The purpose of these specifications was to provide guidelines for the provision of an interaction channel for LMDS networks. These specifications are more suited for systems offering high-speed Internet access to a large number of residential users rather than to a small number of large-scale businesses [2].

7.2.3.2 IEEE Standard

The IEEE organization, which has members across the globe, standardized the IEEE 802.16-2001 standard to target LMDS frequency bands [3]. This standard was published in 2002, and it specifies the air interface for wireless metropolitan-area networks (WMANs) in the 10- to 66-GHz spectrum. The channels are expected to be 20, 25, or 28 MHz wide, which supports data rates in excess of 120 Mbps. The specifications also define a MAC layer that supports a single carrier (SC) based physical layer called WirelessMAN-SC. The physical layer supports both FDD and TDD duplex modes. The FDD mode also supports both full-duplex and half-duplex operation. Overall, the target application provides fixed wireless services that utilize outdoor antennas in LOS conditions in contrast to the handheld devices used by cellular systems.

The uplink physical layer is based on the combination of TDMA and demand assigned multiple access (DAMA), whereas the downlink channel is TDM. The uplink channel is divided into a number of timeslots. The number of slots assigned for various uses (registration, contention, and guard or user traffic) is controlled by the MAC in the base station and may change with time to maximize performance. Each TDMA burst is designed to carry variable-length MAC PDUs. The transmitter randomizes the incoming data; the FEC then encodes it and maps the coded bits to a QPSK, 16-QAM, or 64-QAM constellation. The TDM-based downlink channel carries information for each subscriber station multiplexed onto a single stream of data and received by all sectors within the same sector. The downlink physical layer (PHY) includes a transmission convergence sublayer that inserts a pointer byte at the beginning of the payload to help the receiver identify the beginning of a MAC PDU. The data bits coming from this sublayer are randomized, FEC encoded, and mapped to a QPSK, 16 QAM, or 64-QAM signal constellation. Higher modulation means better spectral efficiency (bits per second per hertz) at the cost of a smaller coverage area. So, in this case 64-QAM will have the highest spectral efficiency, whereas QPSK will provide the largest coverage area.

Besides these standards-based solutions, proprietary LMDS systems are also available. LMDS systems are widely used in Europe for backhaul of wireless networks. However, the LMDS industry has been struggling in the United States with the failure of many companies, providing proof of the importance of scale to the cost effectiveness of fixed wireless solutions.

7.2.4 LMDS Signal Attenuation

One of the major drawbacks of operating at such a high frequency is that signals will be attenuated due to rain, trees, and foliage. Several studies [4–7] have been conducted to understand the levels of attenuation at 28-GHz-plus frequencies. This section briefly presents the results of such studies.

LMDS systems designed to operate in the frequency range of around 30 GHz (10 to 66 GHz in general) are typically limited to operation by means of an unobstructed LOS link due to significant attenuation caused by trees at these frequencies. Operation in the millimeter range imposes some restrictions. Precipitation effects lead to severe attenuation and limit the reliable range of operation to 3 to 5 km depending on the climatic zone and the frequency of operation. On the other hand, because the deployments are fixed, antennas can be designed to be highly directional, capable of beam widths of 2 to 3 degrees at the customer device. As a result, any multipath that may exist in the channel is significantly attenuated by the antenna sidelobes. Therefore, multipath is not a significant channel impairment for LMDS systems in this frequency range. Although multipath is not a major issue for LMDS, it can be reduced by means of BS antennas placed on tall buildings. This fact resulted in the adoption of TDM/TDMA- and FDM/FDMA-based systems rather than other waveform options such as CDM/CDMA, which are better suited to combat severe multipath fading.

However, multipath can occur due to the foliage and the reflection from wet surfaces during rain. The received signal strength during rain varies according to a Rician distribution with a factor inversely proportional to the rain rate. In addition, changes in the electromagnetic properties of the surface or the formation of standing water surfaces during rain may increase the reflected power and result in multipath.

Another signal attenuation factor is the foliage. Swaying foliage in wind causes a significant channel fading at 29.5 GHz. The fading depth, depending on the direction and velocity of wind, tree species, foliage density, humid climate, and very complex leaf structure, is approximately 15 dB. This is because the motion of leaves is comparable to the frequency pattern of 29.5 GHz, thereby having a significant effect on the channel.

As stated earlier, the primary channel impairment that is experienced in these systems is due to rain, which changes relatively slowly in time and is typically addressed by inserting a rain fade margin into the link budget in order to satisfy an availability of 99.9% to 99.99%. Other typical attributes of LMDS systems at these frequencies include cell sizes ranging from 3 to 5 km, the use of sectorized antennas at the BS (typically 90 degrees or better), and aggressive frequency reuse, which allows the reuse of up to 50% to 90% of the available bandwidth from sector to sector. Sectorization is used to increase the capacity of the system via directional antennas of 90 degrees or better.

The bottom line is that the manufacturers and operators of LMDS systems have to pay extra attention to the signal attenuation factor compared to their cellular counterparts.

7.3 Broadband Radio Service Systems

BRS or Multichannel Multipoint Distribution Service (MMDS), as it was formerly called, focuses on the fixed broadband service that operates in the frequency spectrum of <11 GHz. Contrary to the LMDS spectrum, these frequencies do not require LOS and are not get heavily impacted by rain and foliage. The BRS systems are vulnerable to interference and multipath like cellular systems. To address these shortcomings, the IEEE 802.16 LAN/MAN committee endorsed the OFDM (OFDMA) as the key access technology, in contrast to the 3G CDMA systems, for broadband services.

OFDM is a combination of modulation and multiplexing in which the signal is first divided into independent subchannels, then each subchannel is independently modulated, typically using QAM or PSK and then combined to create the OFDM carrier. The subchannels are spaced apart at precise frequencies to provide orthogonality, which prevents the demodulators from seeing frequencies other than their own. OFDM has high resistance against multipath interference and good spectral efficiency, but it tends to suffer from a high peak-to-average power ratio (PAPR). OFDMA is a multiple-access scheme for OFDM systems. It works by assigning a subset of subcarriers to individual users.

IEEE 802.16 addresses the last mile connection by means of WMANs. It defines a MAC layer that supports multiple physical layer specifications that are customized for specific frequency bands. The BRS systems are mainly expected to target the WMAN market in contrast to the WWAN market of cellular systems. It supports PMP and mesh topologies. Though the standard was initially designed for fixed services, efforts are underway to add mobility to the picture.

In 2001, the Worldwide Interoperability for Microwave Access (WiMAX) Forum was created to promote the IEEE 802.16 family of standards. It addresses compatibility and interoperability across multiple suppliers. These topics are not addressed by the current IEEE 802.16 standards. Although WiMAX is only few years old, IEEE 802.16 has been around since the late 1990s to address the broadband market. Although the standard has been endorsed, the FWA market has not taken off, possibly due to the turmoil in the telecommunications industry in the early 2000s [8].

7.3.1 Frequency Spectrum

BWA normally operates in the 2.5-GHz band in North America and in the 3.5-GHz band in the rest of the world. Globally, the BWA systems can operate in both licensed and unlicensed frequency bands. There is a potential for BRS systems to be deployed in the 700-MHz band. On June 10, 2004, the FCC renamed the MDS[1] spectrum as the Broadband Radio Service (BRS) spectrum. The FCC issued a report and ordered the structure of the original MDS (21,550- to 2,162-MHz) and MMDS (2,500- to 266-MHz) bands to be changed, as shown in Figure 7.2. This spectrum rearrangement was done to facilitate the development of broadband wireless data services.

The new band has been divided into three segments: the lower band segment (LBS), midband segment (MBS), and upper band segment (UBS). The LBS and UBS were arranged with the MBS between them to allow duplexer separation for FDD or to provide isolation between TDD access and backhaul systems. The MBS retains the legacy 6-MHz channel widths for analog services and to fulfill educational services. Each band segment consists of four 16.5-MHz blocks where each block has three 5.5-MHz channels.

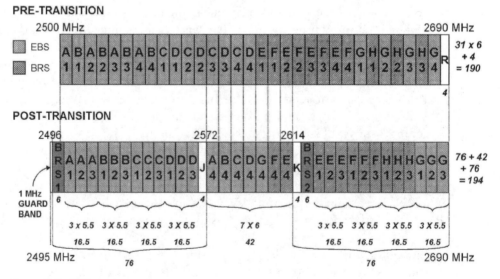

Figure 7.2 BRS spectrum plan.

---

1.   MDS refers to both Multipoint Distribution Service and Multichannel Multipoint Distribution Service (MMDS) licenses.

7.3.1.1   Unlicensed Bands

In unlicensed spectrum, service providers are allowed to offer services without requiring a government license implying exclusive access. The Industry, Science, and Medicine (ISM) bands are typically unlicensed, which in most markets could be used for broadband service. The most common ISM bands are 2.5 and 5.8 GHz. Also, 355 MHz of spectrum is available at 5 GHz under the umbrella of the Unlicensed National Information Infrastructure (UNII).

7.3.1.2   Licensed Bands

As stated earlier, the licensed spectrum is found at 700 MHz, 2.3 GHz, 2.5 GHz, and 3.5 GHz, with the latter two bands currently receiving the most attention.

- *2.5-GHz band:* This band is available in North and Latin Americas and potentially across Europe when the 3G extension band is auctioned in the next few years. It is one of the key bands that could allow mobility for the users.

- *3.5-GHz band:* This band is available everywhere except in the United States. Due to severe RF constraints and regulatory policies on this band, service providers will not be able to offer full mobility to the users.

- *700-MHz band:* Currently, there are no specifications in the standard but there is potential that certain WiMAX supporters will introduce it. This band is currently used by TV broadcasters in North America. It is expected to be vacated in the few years when TV broadcasters switch to digital TV.

- *2.3-GHz band:* This band is currently in use in different countries for specific applications. It does not seem to be a good choice for the United States because the usage in adjacent channels limits the amount of available bandwidth.

7.3.2   Network Topology

This section discusses the three key network topologies for the BWA systems: point-to-point (PTP), point-to-multipoint (PMP), and mesh networks.

7.3.2.1   Point-to-Point Topology

The PTP topology consists of one transmitting and one receiving entity (radio), as shown in Figure 7.3. Because a PTP connection is not shared with multiple users, it delivers more bandwidth to the end user. This topology provides wireless network backhaul, a cost-effective alternative to wired backhaul, which allows service providers to substitute their costly T1 lines with the PTP networks. The radio configuration can be integrated within one box or it can be split with one indoor unit and one outdoor unit.

Figure 7.3   Point-to-point topology.

7.3.2.2   Point-to-Multipoint Topology

The PMP networks allow a single source to serve multiple end users. PMP systems are more economical than PTP links in medium to high traffic density areas. Because many users are sharing the bandwidth in this topology, it is critical to select systems that support efficient resource allocation mechanisms.

PTP systems are used more by residential customers than business users because of less stringent bandwidth requirements. Figure 7.4 shows an example of a PMP configuration in which an access point is transferring information to both residential and business customers.

7.3.2.3   Mesh Topology

In the mesh topology each node has redundant connections to other nodes in the network, as shown in Figure 7.5. The mesh topology can be partial or full. In a partial mesh, each node is connected to at least two other nodes, whereas in a full mesh each node is connected to every other node. A partial mesh in comparison to a full mesh offers less redundancy and greater economics.

One of the key aspects of a mesh network is the routing functionality of its nodes, which allows them to take the best route in communication with other nodes or networks. The failure of a node results in automatic rerouting to bypass the node failure. Another interesting aspect is that a mesh network is independent of the communication system used, and it is applicable to both wired and wireless environments. So mesh networks can utilize BRS or FSO systems.

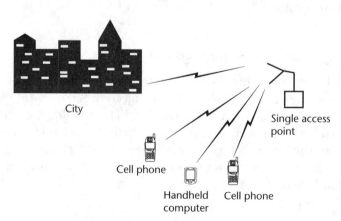

Figure 7.4   Point-to-multipoint topology.

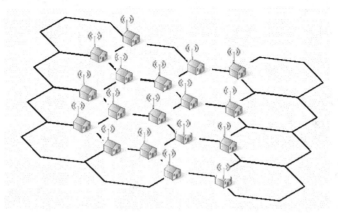

Figure 7.5   Mesh topology.

7.3.3   BWA Standards

The IEEE 802.16 standardization work began in the late 1990s and the first specifications, IEEE 80.16-2001, were completed in April 2002. The IEEE 802.16-2001 standard was designed for 10- to 60-GHz fixed frequencies that require LOS conditions as stated in the earlier LMDS section. The next step was to remove the LOS requirement and add portability and mobilization to the broadband systems. To achieve these tasks, the IEEE 802 committee completed IEEE 802.16-2004 (to remove LOS weakness) and completed IEEE 802.16e[2] (to add mobility) in December 2005. A brief summary of the family of IEEE 802.16 standards follows.

7.3.3.1   IEEE 802.16a-2003

IEEE 802.16a-2003 was as amendment to the IEEE 802.16-2001 standard, introducing modifications in the MAC layer and additional PHY specifications for the licensed and license-exempt band in the 2- to 11-GHz spectrum. In addition to one physical layer supported by IEEE 802.16-2001, IEEE 802.16a-2003 supports two additional physical layers: SC OFDM (256 fast Fourier transform [FFT] only) and OFDMA (2048 FFT only). The target application is fixed/portable wireless service capable of operation in NLOS conditions with outdoor window-mounted antennas and indoor modems. This IEEE 802.16 standard was approved in January 2003 and published in April 2003.

7.3.3.2   IEEE 802.16-2004

The IEEE 802.16-2004 air interface standard was published in October 2004. This standard replaces all previous versions of the IEEE 802.16 standard (including 802.16-2001, 802.16a-2003, 802.16c,[3] and 802.16-REVd[4]). The standard targets fixed wireless services in NLOS conditions with outdoor window-mounted antennas. It usually takes at least 18 to 24 months after the standard is completed for

2.   Only lower-layer specifications were completed.
3.   IEEE 802.16c is a revision of 802.16a, addressing system profile updates.
4.   IEEE 802.16-REVd made modifications to 802.16a.

standards-based equipment and services to be available commercially. Thus, in late 2006 we are starting to see some 802.16-2004 products in the market. The customer-premises equipment (CPE) is not in the handheld form but instead consists of outdoor antennas and indoor modems.

The IEEE 802.16-2004 standard retains all of the major features of the previous 802.16 standards and adds content to improve performance and ease deployment of BWA systems. It allows the telecommunications industry to offer standards-based PMP and mesh topologies in the BRS band along with fixed wireless access applications. The standard supports LAN and MAN networks, LMDS bands, and both licensed and unlicensed bands below 11 GHz. Thus, this standard supports multiple physical layers with advance power management and interference cancellation techniques and multiple antennas for different set of frequencies as shown in Table 7.1. In addition, MAC features such as privacy and automatic repeat request have been included in the standard [9, 10].

Standard Architecture
Next, we briefly define the architecture (Figure 7.6) of this standard and illustrate the layers and SAPs in the next few subsections [9].

MAC Sublayers.   The MAC consists of three sublayers—service-specific convergence sublayer (CS), MAC common part sublayer (MAC CPS), and security sublayer—all of which are discussed in the following subsections.

Service-Specific Convergence Sublayer.   The service-specific CS resides on top of the MAC CPS and utilizes the services provided by the MAC CPS through the MAC SAP. It performs the following functions:

- Accept higher layer PDUs and classify them.
- Deliver CS PDUs to the appropriate MAC SAP and receive CS PDUs from the peer entity.

The standard has defined two CS specifications for mapping services to and from the MAC connections: the asynchronous transfer mode (ATM) CS and the packet CS.

Table 7.1   IEEE 802.16-2004 Air Interface Nomenclature

Frequency Spectrum	Physical Layer Type	Duplexing Mode
10–66 GHz	WirelessMAN-SC	TDD
		FDD
<11 GHz (licensed bands)	WirelessMAN-SCa	TDD
		FDD
<11 GHz (licensed bands)	WirelessMAN-OFDM	TDD
		FDD
<11 GHz (licensed bands)	WirelessMAN-OFDMA	TDD
		FDD
<11 GHz (unlicensed bands)	WirelessHUMAN (high-speed unlicensed MAN)	TDD

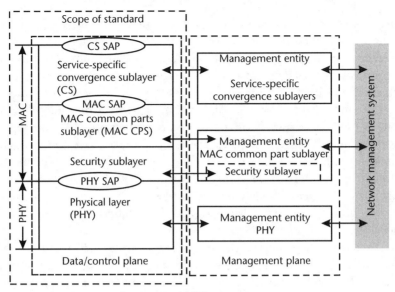

Figure 7.6 IEEE 802.16-2004 reference model [9].

- *ATM CS:* The ATM CS is the logical interface that links the different ATM services with the MAC CPS service access point. The ATM CS accepts ATM cells from the ATM layer and delivers the CS PDUs to the appropriate MAC SAP. It is specifically defined to support the convergence of PDUs generated by the ATM layer protocol of an ATM network. Because ATM cell streams are generated according to the ATM standards, no ATM CS service primitive is required.

- *Packet CS:* The packet CS is used for transport of packet-based protocols such as IP, Point-to-Point Protocol (PPP), and IEEE Std 802.3 (Ethernet). The packet CS performs the following key functions:

  - Classification of the higher layer protocol PDU into the appropriate connection;
  - Delivery of the resulting CS PDU to the MAC SAP associated with the service flow for transport to the peer MAC SAP and receipt of the CS PDU from the peer MAC SAP.

MAC Common Part Sublayer (MAC CPS). The MAC CPS provides the core MAC functionality to support broadband applications. The functionality includes system access, bandwidth allocation, connection establishment, and connection maintenance. The MAC is connection oriented, which means that all services—even those that are inherently connectionless—are mapped to a connection. This provides a mechanism for requesting the core functions, associated QoS, and transporting and routing data to the appropriate convergence sublayer. The connections are referenced with 16-bit connection identifiers (CIDs), where each identifier permits a total of 64K connections within each downlink and uplink channel. The CID serves as a pointer to destination and context information even for connectionless traffic such as IP.

Each subscriber station (SS) or node in terms of the mesh network has a universal 48-bit MAC address as defined in the IEEE 802-2001 standard. It is used during the authentication process by which the network and SS (node) can verify each other. This address also serves as an equipment identifier, since the primary addresses used during operations are the CIDs. At SS initiation in the PMP configuration, three management connections are established between the SS and the BS each with its own level of QoS. The first of these is the basic connection that is used to exchange short, time-critical MAC management messages. The second primary one is used to exchange longer, more delay-tolerant MAC management messages, such as those used for authentication. Further, the secondary management is used for delay-tolerant, standards-based messages such as the Dynamic Host Configuration Protocol (DHCP), Trivial File Transfer Protocol (TFTP), Simple Network Management Protocol (SNMP), and so forth. In addition to these management connections, CIDs are also assigned for higher layer sessions and contracted services.

In terms of mesh network, the candidate nodes are identified via a 16-bit identifier (Node ID), which is transferred in the mesh subheader. The subheader follows the generic MAC header in both unicast and broadcast messages. On the other hand, nodes in the local neighborhood are identified with 8-bit link identifiers (Link IDs). The Link ID is transmitted as part of the CID in the generic MAC header in unicast messages.

MAC Security Sublayer.   The MAC security sublayer provides authentication, secure key exchange, and encryption, which can be summed up as "privacy" across the fixed broadband wireless network. Privacy is achieved by encrypting connections between the base stations and subscriber stations. Encryption is always applied to the MAC PDU payload while the generic MAC header is not encrypted. The privacy features employ an authenticated client/server key management protocol in which the BS acts as a server and controls the distribution of the keying material to the client SS. The privacy mechanisms are strengthened even more by adding X.509 digital-certificate-based SS authentication to its key management protocol (KMP). The KMP also uses Rivest, Shamir, Adelman (RSA) public-key encryption standards to further strengthen the key exchange process between the BS and the SS. Overall, privacy has two component protocols:

- An encapsulation protocol for encrypting packet data across the fixed BWA network. This protocol defines (1) a set of supported cryptographic suites and (2) the rules for applying those algorithms to a MAC PDU payload. A cryptographic suite is the security association's[5] set of methods for data encryption, data authentication, and traffic encryption key exchange. The suite includes algorithms such as Data Encryption Standard (DES), RSA, and Advanced Encryption Standard (AES).
- A KMP provides secure distribution of keying data from BS to SS. It synchronizes keying data between the SS and the BS and BS via KMP enforces conditional access to network services.

---

5.   A security association (SA) is the set of security information shared between a BS and one or more of its client SSs in order to support secure communications across the IEEE 802.16 network.

Physical Layers.   The IEEE 802.16-2004 defines five different types of physical layers as shown in Table 7.1. These PHYs operate in different frequency bands using a variety of modulation and access techniques. This section will briefly highlight the characteristics of these physical layers [9–11].

WirelessMAN-SC PHY.   The SC PHY supports operation in the 10- to 66-GHz band and requires LOS operation as stated in the previous section.

WirelessMAN-SCa PHY.   The WirelessMAN-SCa PHY is also based on single carrier technology, but it was designed to operate in NLOS conditions. It operates in the frequency bands below 11 GHz. It supports channel bandwidths of 3.5, 6, 7, and 20 MHz with data rates in excess of 100 Mbps. Both FDD and TDD duplex modes as well as adaptive modulation with FEC Reed-Solomon coding are supported by this physical layer.

Figure 7.7 illustrates the steps involved in transmission processing. The source data are first randomized, then FEC encoded and mapped to QAM symbols. The QAM symbols are framed within a burst set that introduces additional framing symbols. The symbols within a burst set[6] are multiplexed into a duplex frame, which may contain multiple bursts. The I and Q symbol components are introduced into pulse shaping filters, quadrature modulated up to a carrier frequency, and amplified with power control so that the proper output power is transmitted. The overall operation should be reversed in the receive direction.

WirelessMAN-OFDM PHY.   The WirlessMAN-OFDM PHY is based on OFDM modulation and designed to operate in NLOS conditions and in frequency bands of less than 11 GHz. The OFDM access scheme is mainly aimed at smaller distances (e.g., fixed indoors environment). It supports both PMP and mesh network topologies. The downstream data are transmitted using TDM, whereas the uplink utilizes TDMA. It also supports multiple channel bandwidths: 1.75, 3, 3.5, 5.5, and 7 MHz in the licensed bands with data rates in excess of 20 Mbps. In licensed bands, both FDD and TDD duplexing methods are supported, whereas in license-exempt bands, the duplexing method is TDD.

An OFDM symbol in the frequency domain is made up from subcarriers, the number of which determines the FFT size. There are three subcarrier types, as shown in Figure 7.8:

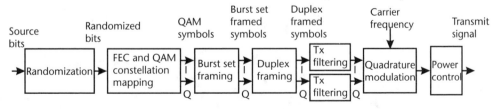

Figure 7.7   WirelessMAN-SCa PHY transmit processing [9].

---

6.   The burst set is a transmission entity consisting of a preamble, one or more concatenated bursts, and a trailing termination sequence.

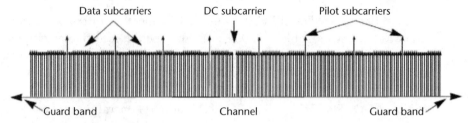

Figure 7.8 OFDM description [9].

- *Data subcarriers:* These are used for data transmission.
- *Pilot subcarriers:* These are employed for several estimation processes.
- *Null subcarriers:* These include nonactive subcarriers and DC subcarriers (center frequency). These are not transmitted and used for guard bands.

In the time domain, the inverse FFT (IFFT) creates the OFDM waveform. When initialized, the SS searches all possible values for the cyclic prefix (CP) until it finds the CP being used by the appropriate BS. The SS uses the same CP in the uplink and the BS should also use the same CP to avoid resynchronization of all the SSs in its area.

WirelessMAN-OFDMA PHY.   This PHY is based on OFDMA modulation with NLOS operation in the frequency bands below 11 GHz. Similar to OFDM, it supports the three types of subcarriers. However, contrary to OFDM, it is subdivided into a subset of subcarriers in which each subset is termed a subchannel. In the downlink, a subchannel is intended for different (group of) receivers; in the uplink, a transmitter is assigned one or more subchannels, and several transmitters can transmit simultaneously. The OFDMA concept is shown in Figure 7.9. In OFDMA, the higher 2,048 FFT is supported. The subchannels in the downlink are used for separating the data into logical streams. These streams employ different modulation and coding schemes to address the needs of subscribers for different channel characteristics. In the uplink, the subchannels are used for multiple access.

It also supports multiple channel bandwidths: 1.25, 3.5, 7, 8.75, 4, 17.5, and 28 MHz in the licensed bands with data rates in excess of 20 Mbps. In licensed bands, both FDD and TDD duplexing methods are supported, whereas in license-exempt bands, the duplexing method is TDD. Through IEEE specifications OFDMA PHY, the commercial products do not support OFDMA.

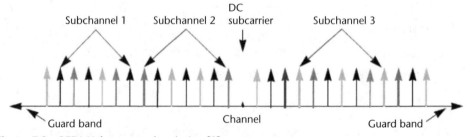

Figure 7.9 OFDMA frequency description [9].

WirelessHUMAN PHY. This high-speed MAN (HUMAN) physical layer uses the licensed-exempt band (i.e., UNII) in the United States or the European Conference of Postal and Telecommunications Administrations (CEPT) band in Europe. These bands fall between 5 and 6 GHz with 10- and 20-MHz-wide channels. The WirelessHUMAN (OFDM) and WirelessHUMAN (OFDMA) technologies support 10-MHz and 10- and 20-MHz channels, respectively.

7.3.3.3 IEEE 802.16e

The IEEE 802.16e standard is an amendment to the IEEE 802.16-2004 standard that allows both fixed and mobile operations in licensed bands below 6 GHz. It is expected to support voice and data at vehicular speeds of up to 120 km/h. Full mobility support is not expected before 2009–2010. The CPEs are expected to be handheld, battery-powered devices. The PHY and MAC layers specifications were standardized in December 2005, and the system profiles were completed in mid 2006, and certification is expected to start in early 2007. The profiles are mainly based on TDD in the 2.5-GHz, 2.3-GHz, and 3.5-GHz bands.

It adds OFDMA 128, 512, and 1024 in addition to 2048 FFT modes to facilitate a scalable OFDMA (S-OFDMA) physical layer. S-OFDMA means that the number of OFDM tones increases based on the quality of the RF signal for a particular user. In addition, 802.16e introduces new header information, which is required for mobility (cell handoffs and so on). It also supports beam-forming and MIMO techniques. The differences between the 802.16e and 802.16-2004 MAC layers would prevent the fixed and mobile versions from working together; that is, 802.16e is not backward compatible with 802.16-2004.

7.3.3.4 IEEE 802.20

In December 2002, the IEEE approved the establishment of 802.20, the Mobile Broadband Wireless Access (MBWA) working group. It was tasked to specify the PHY and MAC layers for mobile BWA systems operating in licensed bands below 3.5 GHz. The working group activities were suspended for a couple of years, and it was revived in 2005 by Qualcomm with their takeover of Flarion Technologies. Qualcomm also submitted a proposal in January 2006 to counter WiMAX. The proposal was an amalgamation of OFDMA and EV-DO technologies. On June 8, 2006, the IEEE-SA Standards Board suspended all the activities of the 802.2 working group. The suspension was effective until October 1, 2006, but it is still ongoing. This suspension was mainly due to two reasons. First, recent activity in the group has become highly controversial, and, secondly, an introductory investigation revealed a lack of transparency and other misconduct in the group.

7.3.3.5 Comparison Matrix

Table 7.2 illustrates the key aspects of the IEEE 802.16 family of standards.

Table 7.2   IEEE 802.16 Family of Standards

Specifications	IEEE 802.16-2001	IEEE 802.16-2004	IEEE 802.16e
Multiple access method	TDMA/FDMA	ODFM/OFDMA (see note)	S-OFDMA
Frequency band	10–66 GHz	10–66 GHz ; licensed and unlicensed bands < 11 GHz	Licensed bands < 6 GHz
Channel bandwidth (MHz)	20, 25, and 28 MHz	1.75/3/3.5/5/5.5/7(OFDM) 1.25/3.5/7/14/28 (OFDMA) (see note)	1.25/2.5/5/10/20 1.75/3/3.5/5.5/7
Mobility	No (fixed)	No (fixed)	Yes
Physical layers	Single carrier	Single carrier OFDM–256 FFT; OFDMA; 2,048 FFT (see note)	Single carrier; OFDM; 256 FFT; OFDMA, (S-OFDMA), 128 FFT, 512 FFT, 1,024 FFT, 2,048 FFT
Modulation	QPSK, 16-QAM, and 64-QAM	BPSK, QPSK, 16-QAM, and 64-QAM	QPSK, 16-QAM, and 64-QAM
Sector throughput (Mbps)	—	4	6
Spectral efficiency (bps)	—	0.6	0.9
Cell radius	1 to 3 miles	1 to 5 miles	1 to 3 miles
Devices supported	Outdoor directional antenna with indoor modems	Outdoor or window-mounted directional antenna with indoor modems	Handheld battery powered devices
Standard completion	April 2002	October 2004	December 2005
Availability	Now	Now	Fixed (2007), partial mobility (2008), full mobility (2009/2010)

Note: Only OFDM 256 FFT is specified in 802.16-2004 profiles.

7.3.4   Key Developments for BRS Adoption

Some of the key developments that are required for the commercialization and widescale adoption of cost-effective, nonproprietary BRS systems are as follows:

- The first is the completion/endorsement of one standard—IEEE 802.16e or IEEE 802.20—that can support both fixed and mobile applications. The endorsement of one BRS standard across the globe would drive down equipment and operational costs and eliminate bifurcation within the broadband industry at large. In other words, the acceptance of one standard will provide economies of scale. Although technically and economically mobility is difficult because of the operation at such frequencies and increased adoption of high-data-rate cellular technologies, the underlying standards that are in works will have to address such limitations.

- It would be prudent to include key features of proprietary solutions (e.g., WiBro, which is deployed in South Korea) in either IEEE 802.16e or IEEE 802.20 to achieve economies of scale.

- IEEE only addresses the PHY and MAC aspects; it does not focus on ensuring interoperability, RF constraints, minimum performance levels, or service layer aspects. Therefore, one of the key goals of the WiMAX forum is to address interoperability and other aspects missing from the standards.

- Though the abundance of spectrum is good news, it will require CPEs to support up to five licensed frequency bands. Initially, it might be worthwhile to focus only on the 2.5- and 3.5-GHz bands. The unlicensed bands will need to mitigate interference and will have to compete very aggressively with other solutions.

- The availability of chipsets, interoperability testing, early technology testing, and market field trials are some of the issues that still need to be addressed.

- Clear differentiation from the cellular networks is needed in terms of handsets, applications, services such as backhaul, and markets. In order for BRS to successfully compete in the mobile environment, it will need to offer better services than those offered by 3G, or it will need to offer the same set of services at a more attractive price.

- Last, but not least, is the justification from the business teams of the service providers for the adoption of BRS systems.

7.4   ETSI Broadband Radio Access Networks

ETSI established a standardization project for BRAN in the spring of 1997. ETSI BRAN has published specifications for three major standard areas:

1. HiperLAN2 (discussed in the next chapter);
2. HIPERACCESS (high-performance radio access), a fixed wireless broadband access network for the 40-GHz band;
3. HIPERMAN (high-performance radio metropolitan-area networks), a fixed wireless access network for 11 GHz and below.

7.4.1   HIPERACCESS

The HIPERACCESS standard area produces standards for broadband multimedia fixed wireless access. The HIPERACCESS specifications allow for a flexible and competitive alternative to wired access networks. HIPERACCESS is targeting high-frequency bands; specifically, it will be optimized for the 40.5- to 43.5-GHz band. BRAN is cooperating closely with IEEE Working Group 802.16 to harmonize the interoperability standards for broadband multimedia fixed wireless access networks.

7.4.2   HIPERMAN

HIPERMAN is an interoperable broadband fixed wireless access system operating at radio frequencies between 2 and 11 GHz (mainly in the 3.5-GHz band). The HIPERMAN standard is designed for FWA provisioning to residences using the basic MAC (DLC and CLs) of the IEEE 802.16-2001 standard. It offers various service categories, QoS, security, fast adaptation of coding, modulation, and transmitting power to propagation conditions and is capable of NLOS operation.

HIPERMAN enables both PMP and mesh network configurations. HIPERMAN also supports both FDD and TDD frequency allocations.

7.5   Free Space Optics

FSO is another means of solving the last mile problem using optical technology. FSO links are also called *fiberless optics* or *optical wireless* transmissions. FSO technologies offer optical capacity but are typically deployed at lengths under a kilometer for reasonable availability. In FSO communications, a narrow beam of light is transmitted by the transmitter through the atmosphere (medium), which is subsequently received at the receiver. Although FSO systems provide an alternative path for broadband wireless applications, the advantages of optical wireless have not yet been fully exploited [12–14].

7.5.1   Basic FSO Architecture

A typical optical wireless communications system consists of three basic components: the transmitter, the propagation channel, and the receiver, as shown in Figure 7.10.

An optical transmitter converts an electrical signal into light. In the transmitter, a light source, which could be a laser or a light-emitting diode (LED), is modulated to carry data. Note that the LED is a lower quality source than the laser, but is preferred over a laser because it is a cheaper alternative. The laser is characterized by its central wavelength, average power, and beam divergence angle. In ideal conditions, its frequency spectrum is sufficiently narrow to allow optical analysis to relate to the central wavelength alone. The beam divergence angle determines the free space power, or how large the laser spot will be at the given distance from the source. The transmitter telescope collimates the beam in the direction of the receiver and determines the beam diameter.

The propagation medium, which in this case is the atmosphere, is not an ideal optical communications channel. It is saturated by molecules and aerosols that

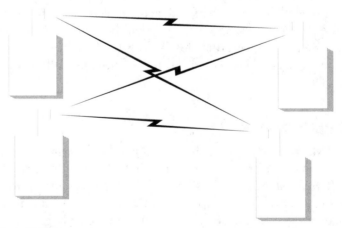

Figure 7.10   Basic FSO schematics.

cause absorption and scattering of light. Other atmospheric factors affecting FSO performance are scintillation, fog, rain, and solar interference.

The light that is emitted from the transmitter propagates through the atmosphere and reaches the optical receiver. At the receiver, a telescope collects the incoming light and focuses it onto the photodetector, which then converts it to electric current. The electrical signal is amplified and processed. A decision-making device determines the nature of the signal according to its amplitude and arrival time. The quality of reception is measured through the BER.

The FSO systems can either use the 1,550-, 780- to 850-, or the 10,0000-nm spectrum. The most common is the 1,550-nm wavelength, which is compatible with erbium-doped fiber amplifier (EDFA) technology, allowing for high-power and high-data-rate systems. Most FSO systems use a simple on-off keying (OOK) modulation scheme to send digital data. In such a system, the presence of light indicates a "1" and the absence of light a "0." OOK is one of the best schemes available to handle atmospheric fluctuations and can be enhanced by additional coding to the channel to reduce the BER.

7.5.2 FSO as an Alternative to RF Communications

Traditionally, the wireless systems that have been implemented to date have been based on RF communications. This is because the telecommunications industry has a vast knowledge base and long experience with RF systems. RF systems offer good coverage and have the ability to penetrate through most building structures. However, these advantageous features become futile when the businesses demand secure and high data rates that require LOS conditions. Additionally, because the radio wireless is a broadcast technology, all users within a cell are sharing a limited number or resources (bandwidth), which diminishes the possibility of providing applications that require very high data rates.

Optical wireless provides a suitable alternative to address these limitations. Also, the RF spectrum is becoming overcrowded, and demand for higher bandwidths is growing rapidly. This LOS communications technology provides a high-data-rate channel exclusively for exchanging information between the two connected parties. Unlike RF systems, FSO does not carry heavy tariffs and licensing fees for the use of radio-frequencies because it does not require any spectrum allocation. Further, the amount of interference between the different channels is very low, even when those are using the same carrier frequency. Optical wireless systems are highly directional, so the probability for the existence of sidelobes is very low compared to RF systems.

There is no doubt that RF wireless systems will continue to serve a large customer base, but the demand for broadband applications will drive the need for alternative choices such as FSO.

7.5.3 FSO Challenges

Although it has many advantages compared to RF systems, FSO does have some technical challenges that need to be addressed. Some of the key ones include maintaining LOS in the presence of building sway, combating light absorption and scattering in different weather conditions, and safeguarding the human eye:

- *Eye safety:* Infrared signals can be hazardous to the human eye. Strict regulations from the American National Standards Institute (ANSI), such as the Z-136 series and the IEC 825 series, help designers protect the eye. The common solution to this issue is to lower the power levels, which in turn requires highly sensitive receivers. These receivers increase the overall complexity of the system. In general, FSO systems operating at 1,550 nm are safer than 780- to 850-nm systems. The 1,550-nm wavelength is too large to cause retina damage because it gets absorbed by the cornea before reaching the eye. This allows about 50 to 60 times higher transmission power with 1,550 nm than 780 to 850 nm from a safety perspective. In addition, the tints normally used by office buildings to filter incoming light are available at the 1,550-nm wavelength.

- *Weather-related issues:* The primary drawback of FSO systems is the degradation in performance during adverse weather conditions. It is worth noting that different weather conditions impact the performance of optical wireless differently. The FSO systems are impacted more by fog and haze than by rain and snow. The main reason behind this is that the radii of these sizable concentrations of scattering particles is on the same scale as the laser wavelength. The scattering, caused by the deflection of incident light from its initial direction, results in the attenuation and distortion of the transmitted signal. Another form of signal attenuation is due to the absorption of laser light energy by the molecules in the atmosphere. It is worth mentioning that different wavelengths could get blocked while others get through by the same set of molecules. Another factor is turbulence, which is due to varying temperature and humidity levels within the propagation medium. Turbulence is responsible for scintillation caused by the fluctuation of the received signal, which can consequently increase the BER of the link.

- *Alignment:* A misalignment between the transmitter and the receiver would lead to additional operational costs. Misalignment is usually caused by the expansion of building frame parts and weak earthquakes. Therefore, alignment needs to be checked periodically to ensure link validity and data transmissions.

- *Light interference:* Infrared signals are prone to interference by natural and artificial light. The solar and fluorescent lights interfere with the infrared signals at the receiver, adversely affecting the performance of the system.

- *Operational issues:* Another challenge for operators is access to weather databases, which is a slow and laborious process. Another key aspect is the site survey for unobstructed LOS conditions.

7.5.4 Mitigation of Challenges to FSO

In the last section, we shed some light on the challenges that are currently faced by FSO systems. In this section, we focus on solutions that could ease the issues on hand from a high level.

- For the protection of the eyes, the industry should abide by the standards. The ANSI Z-136 series and the IEC 825 series would ensure the safety of eyes.
- The optical transmitter power needs to be adjusted to address building swaying. On the one hand, there is a need for higher power for the large divergence angle transmitter that has a high probability of reaching the receiver. On the other hand, low power is sufficient for a smaller divergence angle transmitter that has the probability of missing the receiver. The trade-off between the two provides the optimal values.
- The use of an array of lasers within the transmitter would be helpful for misalignment problems.
- The use of adaptive decision feedback equalizers in the receivers will mitigate the signal scattering and attenuation by reducing the BER.
- The use of modeling tools with embedded databases will be quite useful.

7.6 Millimeter Wave (E-Band)

In October 2003, the FCC announced the availability of frequency bands from 71 to 76 GHz, 81 to 86 GHz, and 92 to 95 GHz for wireless applications. Collectively these frequencies are referred to as the *E-band*, and it is the highest spectrum the FCC has ever licensed to date. The FCC decision provided another opportunity for wireless carriers to compete with the data-carrying capacities of fiber. The millimeter-wave (MMW) spectrum allocation would lead to a broad range of new products and services, including high-speed, point-to-point wireless LANs and broadband Internet access at gigabit data rates and beyond.

Like LMDS and FSO, MMW signals also suffer from atmospheric fluctuations, and unlike RF signals, they have shorter link ranges. This inherent short range is an advantage in that it can be made highly secure and less interfering, especially for dense urban deployments. Another advantage is that MMW antennas are smaller compared to RF for the similar narrow bandwidth. The propagation characteristics at the E-band allow for transmission in excess of a kilometer for 99.999% link availability. At 70 and 80 GHz, the atmospheric attenuation is very similar to the microwave bands of 18 to 38 GHz and less compared to that of the 60-GHz band. It is significantly better than at optical frequencies, where FSO systems can experience approximately 200 dB/km attenuation in thick fog. A comparison of typical 70/80-GHz radio performance against main competing technologies is illustrated in Table 7.3 [15].

Currently, the E-band market is restricted to the United States only; no similar bands are yet available internationally. Despite the potential of being a strong competitor to other broadband technologies, not many companies are actively working to develop systems that can operate in the E-band.

7.7 Summary

A number of broadband wireless access technologies were discussed in this chapter. Some of these do not require LOS conditions and some do, but all are targeted at

Table 7.3 Comparison of Competing Technologies [16]

	70/80 GHz	Microwave Radio (18–38 GHz)	Free Space Optics	Buried Fiber
Data rates	To 10 Gbps	To 322 Mbps	To 10 Gbps	Virtually unlimited
Typical link distances (99.999% availability)	1 km	5 km	200m	Virtually unlimited
Relative cost of installation and ownership	Low	Moderate	Low	High
Installation time	Hours	Days	Hours	Months
Regulatory protection	Yes	Usually	No	Yes

solving the last mile problem. The IEEE 802.16e/802.20 standard still has a long way to go before it becomes a reality. FSO and MMW systems are available, but have to overcome the many challenges associated with atmospheric absorption and operations. In the end, BWA systems have to successfully cross a number of hurdles, as discussed in this chapter, before they can be adopted on a wide scale.

References

[1]   Smith, C., and J. Meyer, *3G Wireless with WiMax and Wi-Fi: 802.16 and 802.11*, New York: McGraw-Hill, 2004.

[2]   "Digital Video Broadcasting (DVB); Interaction Channel for Local Multi-Point Distribution Systems (LMDS)," EN 301 199 (V 1.1.1), European Telecommunications Standards Institute, November 1998.

[3]   "IEEE Standard for Local and Metropolitan Area Networks, Part 16: Air Interface for Fixed Broadband Wireless Access Systems," IEEE Std 802.16-2001, IEEE, April 2002.

[4]   Forester, J., G. Sater, and A. Arunachalam, "LMDS Standards Architectural Issues," *IEEE Wireless Communications and Networking Conf.*, Chicago, IL, September 23–28, 2000, Vol. 3, pp. 1, 590–594.

[5]   Nordbotten, A., "LMDS Systems and Their Application," *IEEE Communications Magazine*, June 2000, pp. 150–154.

[6]   Kajiwara, A., "LMDS Radio Channel Obstructed by Foliage," *IEEE Int. Conf. on Communications*, New Orleans, LA, June 18–22, 2000, pp. 1583–1587.

[7]   Rappaport, S. T., et al., "38-GHz Wide-Band Point-to-Multipoint Measurements Under Different Weather Conditions," *IEEE Communications Letters*, Vol. 4, No. 1, January 2000, pp. 7–8.

[8]   Thelander, W. M., "WiMAX Opportunities, and Challenges in a Wireless World," *Signals Research Group, LLC*, July 2005.

[9]   "IEEE Standard for Local and Metropolitan Area Networks, Part 16: Air Interface for Fixed Broadband Wireless Access Systems," IEEE Std 802.16-2004, June 2004.

[10] Eklund, C., et al., "IEEE Standard 802.16: A Technical Overview of the WirelessMAN™ Air Interface for Broadband Wireless Access," *IEEE Communications Magazine*, June 2002, pp. 98–107.

[11] Koffman, I., and V. Roman, "Broadband Wireless Access Solutions Based on OFDM Access in IEEE 802.16," *IEEE Communications Magazine*, Vol. 40, No. 4, April 2002, pp. 96–103.

[12] Kedar, D., and S. Arnon, "Urban Optical Wireless Communication Network: The Main Challenges and Possible Solutions," *IEEE Optical Communications*, Vol. 42, No. 5, May 2004, pp. 2–7.

[13]   Mahdy, A., and S. J. Deogun, "Wireless Optical Communications: A Survey," *IEEE Wireless Communications and Networking Conf.*, Vol. 5, No. 1, March 2004, pp. 2402–2407.

[14]   Davis, C. C., I. I. Smolyananinov, and D. S. Milner, "Flexible Optical Wireless Links and Networks," *IEEE Communications Magazine*, Vol. 41, No. 3, March 2003, pp. 51–57.

[15]   Wells, J., "Multi-Gigabit Connectivity at 70, 80 and 90 GHz," *Microwave Journal*, Vol. 48, No. 7, July 2005, pp. 128–135.

CHAPTER 8

Fundamentals of WLANs and WPANs

8.1 Introduction

The last few years have witnessed the emergence of wireless local-area networks (WLANs) and wireless personal-area networks (WPANs) in homes, enterprises/ businesses, and public access environments. WLANs and WPANs present two additional choices for mobile broadband users to access their corporate networks and Internet when they are away from their offices. Current WLAN solutions are mainly based on IEEE 802.11 and HIPERLAN standards, whereas WPAN specifications are published by the IEEE 802.15 committee. In addition, some proprietary solutions are also available in the marketplace.

WLANs are envisioned as an alternative to wired LANs, which have high installation and maintenance costs. In public hot spots and homes, WLAN provides high-speed access to the Internet while maintaining an optimal trade-off between range and the data rates. In addition, WLANs can also act as a complement to the 3G cellular networks. The topic of interworking between WLANs and 3G networks will be discussed in the next chapter.

WPANs, on the other hand, are designed mainly to enhance our personal environment within a very short distance, say, a few meters. However, the data rates that are introduced in the 802.15.3 standard are blurring the distinction between the two. WPANs allow us to communicate with the personal devices and provide connections to the outside world. Bluetooth and ultrawideband (UWB) are two commonly known WPANs technologies.

This chapter presents a brief overview of WLAN (IEEE 802.11), HIPERLAN, and WPAN (IEEE 802.15) standards; challenges associated with WLAN and WPAN; and the differences between them. It also elaborates on what the future holds for these networks.

8.2 IEEE 802.11 Standard

The IEEE 802.11 standard is a part of a family of IEEE standards whose goal is to develop specifications for LAN and MAN networks. These standards mainly define the physical layer and MAC sublayer specifications. The IEEE 802.11-based LANs are commonly known as WiFi (wireless fidelity). The WiFi certification is provided by the Wireless Ethernet Compatibility Alliance (WECA). The WiFi components that are certified by WECA are interoperable.

The first IEEE 802.11 standard was approved in 1997 and supported data rates of 1 and 2 Mbps. Two years later the standard was extended to break the 10-Mbps hurdle with 802.11b. Like the original 802.11 standard, it also operates in the unlicensed 2.4-GHz band, but supports higher data rates of 5.4 and 11 Mbps. Besides the original and 802.11b, the IEEE 802.11 committee has also approved 802.11g, 802.11a, 802.11e(for QoS), and 802.11i (for security). Some brief highlights of these standards are provided in the next section [1–3].

8.2.1 WiFi

The four key WiFi standards are 802.11, 802.11a, 802.11b, and 802.11g (Table 8.1). The letter does not imply the sequential availability of the standard. The 802.11b and 802.11g standards are interoperable with each other, whereas 802.11a may coexist in a WLAN environment but its components will not interoperate with 802.11b and 802.11g. Brief definitions of some WiFi standards and extensions are as follows:

- *Original 802.11:* As stated earlier the original 802.11 operates in the 2.4-GHz ISM band and supports 1- and 2-Mbps data rates. Its MAC functionality is based on carrier sense multiple access with collision avoidance (CSMA/CA). It also defines three physical layers: two in the 2.4-GHz band using either frequency-hopping spread spectrum (FSSS) or direct-sequence spread spectrum (DSSS), and the third is based on infrared.
- *IEEE 802.11b:* This is one of the most successful WLAN standards to date and the one most likely found in public hot spots. The specification introduces complementary code keying (CCK) to support data rates of 5.4 and 11 Mbps in the 2.4-GHz band. In addition, it is backward compatible with 802.11.
- *IEEE 802.11a:* Following 802.11b was 802.11a, which specified a new physical layer in the 5-GHz UNII band. The standard uses OFDM instead of FHSS or DSSS and can operate at 54 Mbps. It also supports adaptive modulation and coding, which is also known as link adaptation. AMC allows the modulation schemes to vary from BPSK and 64-QAM according to the RF conditions to support the data rates ranging from 6 to 54 Mbps.
- *IEEE 802.11g:* This standard also supports OFDM and a peak data rate of 54 Mbps. The standard employs FSS and is backward compatible with 802.11b. It operates in the 2.4-GHz ISM band.

Table 8.1 WLAN Standards

Features	802.11	802.11b	802.11a	802.11g
Maximum data rate (Mbps)	1 or 2	11	54	54
Number of overlapping channels	—	14	12	14
Frequency allocation (GHz)	2.4	2.4	5	2.4
Spread spectrum	DSSS/FSSS	DSSS/FSSS	OFDM	OFDM

- *IEEE 802.11e:* This standard is not an air interface standard; instead, it adds QoS features and multimedia support to the existing 802.11b and 802.11a WLAN standards, and thus it is backward compatible with both.

- *IEEE 802.11i:* Like 802.11e, 802.11i is not an interface standard; instead, it adds Advanced Encryption Standard (AES) security protocol to the 802.11 standard. AES is a much better security protocol than the existing Wired Equivalency Privacy (WEP) algorithm.

- *IEEE 802.11n:* The 802.11n standard is expected to provide four to eight times higher data rates than the existing 802.11 products. The standard is expected to be completed by mid-2007.

8.2.2   802.11 Standard Overview

The 802.11 standard defines the physical layer and MAC sublayer for the WLAN systems. The standard makes use of the existing 802.2 logical link control function like other 802 LANs, which allows easy bridging from wireless to IEEE wired networks. The WLAN systems mainly consist of a station (STA), such as a cellular terminal; an access point (AP), such as a base station; and a portal. A portal is a particular AP that connects the wireless LANs with wired LANs.

The basic service set (BSS) is the primary building block of the IEEE 802.11 WLAN architecture. It consists of a set of stations and an AP. The geographical area covered by the BSS is known as the basis service area (BSA) or cell. In its simplest form, the BSS is comprised of two or more STAs that can communicate with each other directly. This mode of operation is often referred to as an ad hoc network because it is created without the assistance of any infrastructure network. The formal name of this mode is independent BSS (IBSS) (Figure 8.1) [4].

The second type of BSS is an infrastructure BSS, which, in contrast to an ad hoc network, is established to provide wireless users with specific services and a range of extensions. Larger geographical areas can be covered with multiple BSSs, such as multiple cells. The backbone network used for the interconnection of a group of

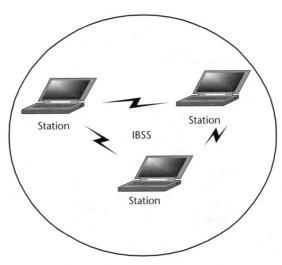

Figure 8.1   Ad hoc network (IBSS).

BSSs is the distribution system (DS). The interconnection of some BSSs and a DS is identified as a single wireless network called an extended service set (ESS), which is shown in Figure 8.2.

The DS is considered to be a backbone network that is responsible for MAC-level transport of MAC service data units (MSDUs). It is worth noting that the DS, as specified by IEEE 802.11, is implementation independent. Therefore, the DS could be a wired IEEE 802.3 Ethernet LAN, fiber distributed data interface (FDDI), MAN, or another IEEE 802.11 wireless medium. The DS is solely used as a transport backbone to transfer packets between different BSSs in the ESS. An ESS can also provide gateway access for wireless users into a wired network such as the Internet via a portal. The portal is a logical entity that specifies the integration point on the DS where the IEEE 802.11 network integrates with a non-IEEE 802.11 network. If the network is an IEEE 802.X (Figure 8.2), the portal incorporates functions that are analogous to a bridge; that is, it provides range extension and the translation between different frame formats.

8.2.2.1   Physical Layer

The standard defines the following three physical layers to support a unique MAC layer defined in the next section [4–7]:

- FSSS in the 2.4-GHz ISM band;
- DSSS in the 2.4-GHz ISM band;
- Infrared (IR).

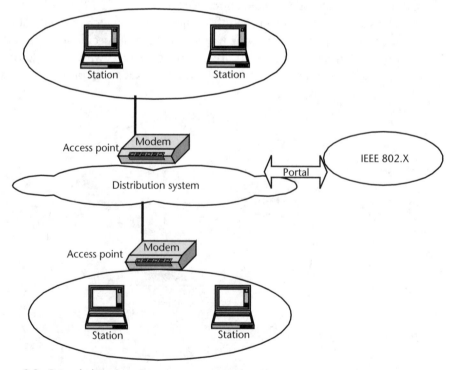

Figure 8.2   Extended service set.

In the FHSS, the 2.4-GHz band is divided into seventy-nine 1-MHz sub-channels. The 1-Mbps data rate is modulated using two-level Gaussian frequency shift keying (2GKSF), whereas the 2-Mbps rate uses four-level GFSK where 2 bits are encoded at a time using four frequencies.

Contrary to FHSS, the DSSS techniques divide the 2.4-GHz band into eleven, thirteen, or fourteen 22-MHz channels. In addition, the spreading of data symbols is done using an 11-chip Barker sequence. In DSS, the 1-Mbps rate is encoded using differential binary phase shift keying (DBPSK), and the 2-Mbps rate uses differential quadrature phase shift keying (DQPSK).

The IR specification identifies a wavelength range from 850 to 950 nm. The IR band is specifically designed for indoor use only and operates with nondirected transmissions; that is, via diffuse transmissions—the receiver and transmitter do not have to be aimed at each other and do not need a clear LOS. The transmission distance is limited to the range of 10 to 20m and the signal is contained by walls and windows. The encoding of the 1-Mbps data rate is performed using 16-Pulse Position Modulation (16-PPM), where 4 data bits are mapped to 16 coded bits for transmission. The 2-Mbps data rate is performed using 4-PPM modulation, where 2 data bits are mapped to 4 coded bits for transmission. The purpose of having PPM for the two different data rates is to ensure that the basic pulse is the same at both data rates, which minimizes the additional complexity introduced by the 2-Mbps data rate. The PPM signals at 1 and 2 Mbps are represented in Figure 8.3. The duration of each pulse is 250 ns, and the peak optical power is 2W. Therefore, the average optical power is 250 mW at 2 Mbps and 125 mW at 1 Mbps.

Table 8.2 recaps some of the differences between the physical layers that were discussed in the preceding paragraphs.

8.2.2.2   MAC Sublayer

The MAC sublayer is responsible for the channel allocation procedures, PDU addressing, frame formatting, error checking and fragmentation, and reassembly of frames. The standard defines a single MAC sublayer for all three aforementioned physical layers. The MAC supports both asynchronous and time-bounded traffic

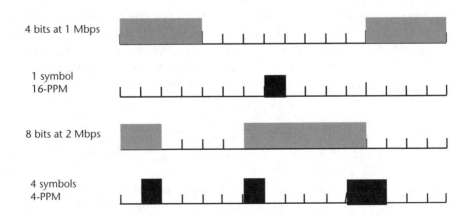

Figure 8.3   PPM signals at 1 and 2 Mbps [6].

Table 8.2 Physical Layer Specifications

Feature	FSSS	DSSS	IR
Spectrum	2.4-GHz ISM band	2.4-GHz ISM band	Diffuse infrared ($\lambda = 850$–950 nm)
Subchannels	79, each 1 MHz	11, 13, or 14, each 22 MHz	N/A
Data rate	1 and 2 Mbps	1 and 2 Mbps	1 and 2 Mbps
Modulation	2GFSK, 4GFSK	DBPSK, DQPSK	4-PPM 16-PPM

through the use of a distributed coordination function (DCF) and point coordination function (PCF), respectively, and power-saving mechanisms.

Distributed Coordination Function (DCF)

The DCF is the primary access method, and it is used to support asynchronous data transfer on a best effort basis. The DCF operates solely in the ad hoc network, and either operates solely or coexists with the PCF in an infrastructure network. Unlike 802.3 Ethernet LANs, which use carrier sense multiple access with collision detection (CSMA/CD), the MAC is based on CSMA/CA, because in 802.11 collision detection is not possible due to the near far problem.[1] In the CSMA/CA protocol, the random back-off time is distributed according to a uniform distribution (in discrete slot times), where the maximum extent of the uniform range is called the contention window (CW). The CW parameter (i.e., the range of this uniform distribution) is doubled—up to a maximum limit—each time a frame transmission is unsuccessful, as determined by the absence of an ACK frame. This exponential back-off mechanism helps reduce collisions in response to an increasing number of contenting stations.

Another problem related to the WLAN MAC is the hidden node challenge, where two stations on the opposite sides of an access point can both hear from an AP but not from each other, usually because of distance or an obstruction. To address this issue, WLAN specifies an optional protocol request to send/clear to send (RTS/CTS) at the MAC sublayer. When this protocol is active, a sending STA transmits a RTS and waits for the AP to reply with CTS. The CTS causes the delay of any intended transmissions to the AP, allowing the sending station to transmit and receive a packet acknowledgment without any chance of collision. Thus, CTS grants the requesting station permission to transmit while at the same time notifying all stations within radio range not to initiate any transmissions for a given time, which is called the net allocation vector (NAV). This feature is mainly used for large packets because it adds additional overhead to the network.

Point Coordination Function

To support time-bounded services, the 802.11 standard specifies the optional use of PCF in which an AP (or PCF station) has priority control of the medium. In other

---

1. Near far problem: Consider transmitters, one close to the receiver and one far away. If both transmitters transmit at the same power, then due to the inverse square law, the receiver will receive more power from the nearer transmitter. That will make it very difficult to hear from the far away transmitter. If the closer one transmits at a level higher than the far away one, then the farther one will simply not be heard by the receiver.

words, when the PCF is active, the PCF station allows only a single station in each cell to have priority access to the medium at any one time. Having silenced all stations, the PCF station can then allow a given station to have contention-free access through the use of an optional polling frame that is sent by the PCF station.

8.2.2.3   MAC and QoS Enhancements in 802.11e

With DCF all of the stations and data flows have the same priority to access the medium, thus there is no QoS in the original 802.11 standard. The DCF is unsuitable for time-sensitive communications, such as voice, and multimedia applications, such as video streaming, because in DCF a station might have to wait an arbitrarily long time to transmit a frame.

To support QoS, the 802.11e adds a new channel access function called the hybrid coordination function (HCF) that includes both controlled contention-free and contention-based channel access schemes. The HCF's contention-based channel access method is called enhanced DCF (EDCF). EDCF provides differentiated distributed access to the wireless medium for eight priorities of stations. It also defines the access category mechanism that provides support for the priorities at the stations. QoS support is realized with the introduction of traffic categories. Each station may have up to four access categories to support eight user priorities. One or more user priorities is assigned to one access category. A station accesses the medium based on the access category of the frame to be transmitted. The mapping between priorities to access categories is defined in Table 8.3 [8, 9].

8.3   HIPERLAN

HIPERLAN is a European alternative to the IEEE 802.11 family of standards. The HIPERLAN standard is defined by ETSI under the umbrella of the Broadband Radio Access Networks (BRAN) project. In parallel to the activities in Europe and the United States, the Multimedia Mobile Access Communications (MMAC) Promotion Association along with the Association of Radio Industries and Broadcasting (ARIB) in Japan started to develop high-speed radio access systems for business

Table 8.3   Mapping Between Priorities and Access Categories

Priorities	Access Category	Traffic Type
1	0	Best effort
2	0	Best effort
0	0	Best effort
3	1	Video probe
4	2	Video
5	2	Video
6	3	Voice
7	3	Voice

and home applications at 5 GHz. One of these systems for business applications in corporate and public networks, HiSWAN (high-speed WAN), has been aligned with HIPERLAN2 [2, 10].

The HIPERLAN family of standards has four alternatives, as shown in Table 8.4, with the most common one being HIPERLAN2. The focus of this section is on HIPERLAN2. We briefly present the key aspects of the standard; the details can be found in the HIPERLAN2 standard [11].

Before starting the standardization work on HIPERLAN2, the BRAN group developed the HIPERLAN1 standard to support asynchronous and synchronous traffic. The standardization work started in 1991 and the standard was approved in 1996. It is similar to 802.11a and covers the physical layer and MAC sublayer like the other 802.11 standards. However, there is a new sublayer called channel access and control (CAC) that deals with the access requests to the channels. The CAC employs an access method called elimination-yield nonpreemptive multiple access mechanism (EY-NPMA), which constitutes a kind of CSMA/CA that splits the procedure into three steps: priority resolution, elimination, and yield. Some of the key features of HIPERLAN1 are as follows:

- Short range—50m;
- Low mobility—1.4 m/s;
- Supports asynchronous and synchronous traffic;
- Sound at 32 Kbps, 10-ns latency;
- Video at 2 Mbps, 100-ns latency;
- Data at 10 Mbps.

8.3.1 HIPERLAN2

The HIPERLAN2 specification was completed in 2002 and was designed to provide fast wireless connections to UMTS, ATM, and IP networks. It uses the 5-GHz band, supports data rates up to 54 Mbps, and has mobility of 10 m/s. HIPERLAN2 also supports QoS by allowing radio bearers to be set up and treated differently by the AP during transmission.

HIPERLAN2 supports two modes of operation—centralized and direct. The centralized mode is used in the cellular networking topology, where each radio cell is covered by an AP covering a certain geographical area. In this mode, a mobile terminal (MT) communicates with other terminals or the network via APs. The

Table 8.4 The BRAN Family

HIPERLAN Type 1	HIPERLAN Type 2	HIPERLAN Type 3 (HIPERACESS)	HIPERLAN Type 4 (HIPERLINK)
WLAN	Wireless IP, ATM, and UMTS short-range access	Wireless IP, ATM remote access	Wireless broadband Interconnect
MAC	DLC	DLC	DLC
PHY (5 GHz, 10 Mbps)	PHY (5 GHz, 54 Mbps)	PHY (various bands, 25 Mbps)	PHY (17 GHz, 155 Mbps)

centralized mode is mainly used in business applications, both indoor and outdoor, when an area much larger than a radio cell has to be covered. The direct mode is used in the ad hoc networking topology typically in home environments, where a radio cell covers the whole serving area. In this mode, the mobile terminals in a single-cell home "network" can directly exchange data. In both cases, access to the medium as well as assignment of radio resources to terminals is controlled by the AP.

We now look at the protocol stack of HIPERLAN2, which consists of three layers: the physical (PHY) layer, the data link control (DLC) layer, and the convergence layer (CL), which is part of DLC. Each of the layers is divided into a user plane and control plane, where the U-plane includes functions such as transmission of traffic over the established user connections, and the C-plane coordinates functions related to the control, establishment, release, and modification of the connections. The protocol reference model of HIPERLAN2 is shown in Figure 8.4 [2, 10, 11].

8.3.1.1 Physical Layer

The physical layer provides data communications via a baseband modem and associated RF circuitry. The basic units to be transmitted via PHY are bursts with variable length. Each burst consists of a preamble and a data field. The OFDM is the selected modulation scheme due to its very good performance on highly dispersive channels. Some main features of the physical layer are listed in Table 8.5.

The channel spacing is 20 MHz with subcarrier per channel spacing of 12.5 kHz. Each channel has 52 subcarriers, out of which 48 subcarriers carry actual data and 4 subcarriers are pilots, which facilitate coherent demodulation. The duration of cyclix prefix (guard interval) is equal to a maximum of 800 ns and a minimum of 400 ns. The shorter guard interval can be used for short-range indoor applications.

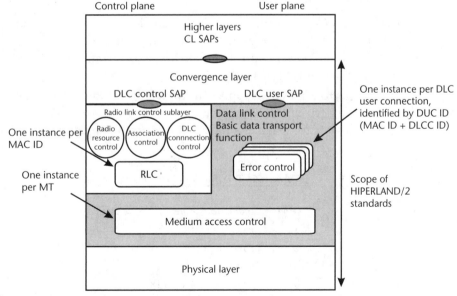

Figure 8.4 HIPERLAN2 protocol reference model [11]. (*Source:* © European Telecommunications Standards Institute 2005. © European Broadcasting Union 2005. Further use, modification, redistribution is strictly prohibited. ETSI standards are available from http://pda.etsi.org/pda/ and http://www.etsi.org/services_products/freestandard/home.htm.)

Table 8.5   HIPERLAN2 PHY Parameters

Attribute	Value
Spectrum	5 GHz
Channel spacing	20 MHz
Subcarriers per channel	52 (48 to carry data and 4 as pilots)
Cyclic prefix	Maximum: 800 ns; minimum: 400 ns
Forward error control	Convolutional code
Modulation	BPSK, QPSK, 16-QAM, 64-QAM
Data rate	6–54 Mbps

The physical layer also supports link adaptation with various code rates and modulation schemes. The seven modes of physical layer support four key modulation schemes—BPSK, QPSK, 16-QAM, and 64-QAM—as shown in Table 8.6.

8.3.1.2   Data Link Control Layer

The DLC represents the logical link between an AP and its associated mobile terminals. It implements a service policy that considers QoS for each connection, channel quality, number of terminal devices, and medium sharing with other access networks operating in the same area. It supports both logical and transport channels such as 3G mobile technologies.

The DLC layer consists of the error control (EC), the RLC, and the MAC sublayers. The DLC layer is divided into data transport and control functions. The data transport handles the data packets arriving from the higher layer via the user service access point (U-SAP). It also contains the EC that performs an ARQ protocol. The control part contains the RLC function as explained here:

- *RLC:* The RLC performs three main functions: association control function (ACF) for authentication and encryption, radio resource control (RRC) for handover and power control, and provision of a transport service to the DLC connection control (DCC).
- *Error control:* The EC supports *acknowledged mode,* for reliable transmissions using retransmissions; *repetition mode,* also for reliable transmission, but without a feedback channel; and *unacknowledged mode,* for unreliable,

Table 8.6   PHY Modes of HIPERLAN2

Mode	Modulation	Code Rate	Physical Layer Bit Rate (Mbps)
1	BPSK	1/2	6
2	BPSK	3/4	9
3	QPSK	1/2	12
4	QPSK	3/4	18
5	16-QAM	9/16	27
6	16-QAM	3/4	36
7	64-QAM	3/4	54

low-latency transmissions. Unicast data can be sent using either the acknowledged or unacknowledged mode. Broadcast services can be supported by either the repetition or unacknowledged mode. Multicast services can be sent in the unacknowledged mode or be multiplexed onto already existing unicast transmissions.

- *MAC:* The MAC protocol is based on TDD and dynamic TDMA. The MAC basic frame structure has a fixed duration of 2 ms and comprises fields for broadcast control, frame control, access feedback control, data transmission in downlink as well as uplink, and random access. The MAC protocol and the frame structure also support multibeam antennas with up to eight beams to improve the link budget.

8.3.1.3   Convergence Layer

The CL is also divided into a data transport and a control part. The two key functions of CL are first, to adapt service requests from higher layers to the service offered by DLC and second, to convert the higher layer packets with fixed or variable size into a fixed-size service data unit (SDU) that is used within the DLC. The convergence layer thus maps the incoming data onto different bearers of the DLC. The CL could be either cell-based for variable-length packets (e.g., Ethernet) or packet-based for fixed-size packets (e.g., ATM). The functionality of padding, segmentation into fixed-size DLC SDUs, and respective reassembly is one key issue that makes it possible to implement DLC and PHY layers that are core network independent. For example, a higher layer packet such as an Ethernet packet is mapped onto one or more DLC SDUs by padding and segmentation. In addition, UMTS, Ethernet, and SSCS are defined to perform the appropriate service adaptation.

8.4   IEEE 802.11a Versus HIPERLAN2

There are many similarities and some differences between 802.11a and HIPERLAN2, which both operate in the 5-GHz frequency. The physical layers of IEEE 802.11a and HIPERLAN2 are very similar, both using OFDM and supporting multiple modes and raw data rates of up to 54 Mbps. On the other hand, HIPERLAN/2 and 802.11a use different training sequences in the preamble. The training symbols used for channel estimation are the same, but the sequences provided for time and frequency synchronization are different. Decoding of the convolutional code is typically implemented by means of a Viterbi decoder [12]. At the same time, the actual data rates and data throughputs are highly dependent on the type of MAC sublayer that is used.

The main differences between the IEEE 802.11a and HIPERLAN2 standards occur at the MAC layer. In 802.11a, the MAC is based on CSMA/CA, whereas it uses TDMA in HIPERLAN2. Another difference between the two standards is the length of the packets employed: HIPERLAN2 employs fixed-length packets; 802.11a supports variable-length packets. A comparison between the two is shown in Table 8.7.

Table 8.7   802.11a Versus HIPERLAN2

Features	802.11a	HIPERLAN2
Frequency band	5 GHz	5 GHz
Maximum data rate	54 Mbps	54 Mbps
Spread spectrum	OFDM	OFDM
Packet size	Variable	Fixed
MAC	CSMA/CA	TDMA

It is worthwhile to point out that 802.11a is generally not implemented in Europe due to contention with HIPERLAN2; 802.11b/g systems are preferred because they operate in a different frequency band (2.4 GHz).

8.5   Challenges with Wireless Local-Area Networks

To make WLANs more effective for the everyday users, some enhancements are required. These enhancements can be broadly divided into the following categories, which are both technical and business related [13, 14]:

- *Ease of use:* WLANs should be user friendly so that even novices can handle them easily. The hassle-free configuration process will add to the popularity of WLANs.

- *Interference:* Because both ISM and UNII are unlicensed, interference is a major factor affecting the performance of WLANs. Interference could be caused by microwave ovens, cordless phones, Bluetooth, and so forth.

- *Security:* Security is another major task that needs to be addressed, especially how to handle security protocols between the WLANs and wired LANs, 3G, WPAN, and so forth. IEEE 802.11i defines security measures only for the WLANs, but does not address the aspects of secure communication between WLANs and wired LANs.

- *Mobility:* Providing WLAN mobile capabilities within a building is relatively simple. However, extending it to hot spots via integration with cellular wireless wide-area networks (WWANs) is not an easy task. The interworking between WLANs and WWANs is addressed in the next chapter.

- *Challenges with the UNII band:* Operators have to compensate for an additional 6.38 dB of free space loss at 5 GHz compared to 2.4 GHz. This will result in more APs (at an additional cost) at 5 GHz to cover the same coverage area. Another issue in designing products in the UNII band is that the allocated frequencies are separated. The upper band is at least 400 MHz away from the nearest middle band channel. This presents significant challenges for RF front-end designs if the devices are intended to cover the entire UNII band.

- *Competition from EV-DO and HSDPA:* The availability of EV-DO and HSDPA air cards (network interface cards) is becoming a major challenge for WLANs operators. An EV-DO or a HSDPA subscriber can access the Internet via his or her associated air card in the same way that WLAN users do. This threat from the 3G technologies needs to be carefully addressed by WLAN operators.

- *Network management:* The operations and management of WLANs is a big challenge, especially in hot spots. WLANs are frequently laid over an existing infrastructure with only rudimentary tools for managing day-to-day operations compared to 3G systems.

8.6   Wireless Personal-Area Networks

The WPANs consist of those technologies that allow us to communicate with every-day devices (such as laptops, printers, and fax machines) and with the outside world via wireless links for a very short distance. The distance is usually on the order of a few meters. WPAN enhances our personal experience whether at work or at home by providing networking capabilities with the devices that are in our surrounding space. WPANs will primarily serve as a replacement for the interconnect cables between the above-mentioned and other personal devices.

WPAN activity started with the advent of Bluetooth in 1994 by L. M. Ericsson of Sweden [15]. In the late 1990s, the IEEE took a step forward and started the standardization work for WPANs. The WPANs standards are covered by the IEEE 802 Committee and, more specifically, by the IEEE 802.15 Working Group. The IEEE 802.15.1-2002 publication provides an additional resource for those who implement Bluetooth technology. Currently, the 802.15 Working Group is focusing on defining the next generation of WPANs that implement technologies such as UWB.

In this section, we will provide an overview of IEEE 802.15 standards, Bluetooth and UWB technologies, applications, and challenges associated with WPANs.

8.6.1   IEEE 802.15 Overview

The IEEE 802.15 task group has been asked to define the WPANs standards to enrich our personal networking experience. The 802.15 working group has so far defined four WPAN standards with the help of four task groups, which are briefly defined in this section:

- *IEEE 802.15.1:* This was published in 2002, and it deals with Bluetooth. It includes a PHY and MAC layer specification adapted from Bluetooth 1.1 specifications. A brief overview of Bluetooth technology is provided in the next section.
- *IEEE 802.15.2:* This specified a set of recommended practices to facilitate the coexistence of WPANs (802.15) and WLANs (802.11). The standard was published in August 2003. Task Group 2 is now hibernating until further notice. The interworking scenarios will be discussed in the next chapter.
- *IEEE 802.15.3:* Task Group 3 deals with high-rate WPAN standards (20 Mbps or higher). It will also provide low-power, low-cost solutions to address the needs of portable consumer digital imaging and multimedia applications. The high-rate WPANs operate in the unlicensed 2.4-GHz band with data rates

of 55 Mbps or less. The 802.15.3 standard was approved in September 2003 and some of the main characteristics of the high-rate WPAN standard are as follows [16]:

- Supports multiple data rates: 11, 22, 33, 44, and 55 Mbps;
- Allows ad hoc peer-to-peer networking;
- Supports QoS isochronous protocol;
- Provides advanced power management to save battery power;
- Develops low-cost and complex MAC and PHY implementations optimized for short-range (less than 10m) communications;
- Supports requirements for portable consumer imaging and multimedia applications.

The 802.15.3 physical layer operates in the unlicensed frequency band between 2.4 and 2.4835 GHz and employs the same symbol rate, 11 Mbaud, as used in the 802.11b systems. The PHY supports five different modulation schemes: QPSK at 22 Mbps, trellis-coded QPSK at 11 Mbps, and 16-, 32-, and 64-QAM at 33, 44, and 55 Mbps, respectively. The 802.15.3 signals occupy a bandwidth of 15 MHz, which allows a maximum of four fixed channels in the unlicensed 2.4-GHz band. The transmitting power level complies with the FCC 15.249 rules with a target value of 0 dBm. The MAC supports CSMA/CA like 802.11 and provides QoS.

- *IEEE 802.15.4:* This standard, published in May 2003, specifies a low data rate, but a very long battery life (months or even perhaps years). Since March 2004, Task Group 4 has been hibernating with the creation of Task Group 4b. The new Task Group 4b is chartered to clarify and enhance specific parts of the Task Group 4 standard. The ZigBee set of communication protocols is based on the specification produced by this task group. Here are some of the features of the 802.15.4 standard:
  - Supports data rates of 250, 40, and 20 Kbps;
  - Supports 16-bit short and 64-bit IEEE addressing modes;
  - Supports CSMA-CA channel access and ensures low power consumption;
  - Physical layer supports 16 channels in the 2.4-GHz ISM band, 10 channels in the 915-MHz band, and 1 channel in the 868-MHz band.

8.6.2 Bluetooth

The name *Bluetooth* [2, 5, 15, 17, 18] comes from the Danish king Haralad Blatand (Bluetooth) who was credited with uniting the Scandinavian people during the 10th century. The Bluetooth Special Interest Group (SIG) was founded in February 1998 by Ericsson, IBM, Intel, Nokia, and Toshiba to develop specifications for Bluetooth. The first set of specifications, version 1.0A, was completed in 1999 and enabled short-range wireless connectivity. Later, IEEE 802.15.1 adopted version 1.1 of the SIG specifications.

Bluetooth enables users to connect to a wide range of computing and telecommunications devices without the use of proprietary cables. One connection capability, the cable solution, was complicated because a specific cable was often needed to hook up to a particular device. The infrared solution also had its problems: it required LOS conditions. The Bluetooth standard was developed to solve all such problems.

The Bluetooth system operates in the 2.4-GHz ISM band in most parts of the world. The frequency range is 2.4 to 2.4835 GHz, and it applies/allows frequency hopping over 79 carriers. Besides Bluetooth, several other devices including garage door openers, baby monitors, microwave ovens, and cordless phones also operate in the ISM band. Thus, Bluetooth is susceptible to strong interference from these devices. Bluetooth supports simultaneous voice and data sessions—supporting one asynchronous data channel and up to three synchronous voice channels, or one channel supporting both voice and data.

8.6.3   Next Generation of WPANs

In February 2002, the FCC approved 7.5 GHz of bandwidth for UWB (the next generation of WPAN) in the frequency range of 3.1 to 10.6 GHz. This ruling was preceded by another in April 2002 when the FCC allowed the commercial use of wireless devices based on UWB. Until that time, UWB technology had only been used by the military and government.

UWB spreads the signals (data) over very large bandwidths instead of one narrow bandwidth. It requires very little power to transmit large chunks of data over a short distance. UWB devices use power in the range of 50 to 70 mW, which is considerably lower than that used by cell phones. However, the width of its transmission band is of concern to industry at large because there are many sources of interference in this band, including the harmonics of microwave ovens and ignition coils, and the harmonics of mobile phone transmitters in the 800-MHz band. UWB currently supports 100-Mbps data rates with the potential of 1 Gbps in the future. UWB has the potential to be a disruptive wireless technology—unlike other radio technologies that use one frequency and require considerable power for their transmission, UWB transmits data over a wide spectrum and consumes minimal power [19].

8.6.3.1   UWB—Competing Forces

The IEEE 802.15.3a Task Group that had been working for the last few years to define the next generation of WPANs dissipated in January 2006. The working group failed to decide between the two competing proposals, namely, multiband orthogonal frequency-division multiplexing (MB-OFDM) UWB, supported by the WiMedia Alliance, and direct-sequence UWB (DS-UWB), supported by the UWB Forum. The MB-OFDM solution has the backing of companies such as Intel, Hewlett-Packard, Microsoft, Nokia, and Texas Instruments. The DS-UWB proposal has the support of Freescale Semiconductor, Fujitsu, Siemens, ZTE, and so forth. The dissolution of the working group will likely result in the creation of two distinct UWB solutions (Figure 8.5).

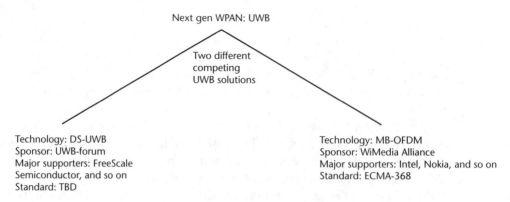

Figure 8.5   UWB-DS versus MB-OFDM.

WiMedia Alliance
On December 8, 2005, ECMA International released two international ISO-based specifications (ECMA-368 and ECMA-369) for UWB technology based on the WiMedia MB-OFDM–based radio platform. The ECMA-368 (high-rate ultra-wideband PHY and MAC standard) specification defines the PHY and MAC sublayer for a high-speed short-range wireless network, utilizing all or part of the spectrum between 3.1 and 10.6 GHz and supporting data rates of up to 480 Mbps. A MB-OFDM scheme is specified to transmit information that uses a total of 110 subcarriers (100 data carriers and 10 guard carriers) per band. In addition, 12 pilot subcarriers allow for coherent detection. Frequency-domain spreading, time-domain spreading, and FEC coding are provided to perform under a variety of channel conditions.

The MAC sublayer is designed to enable mobility, such that a group of devices may continue communicating while merging or splitting from other groups of devices. To maximize flexibility, the functionality of this MAC is distributed among devices to avoid interference and ensure QoS. The MAC sublayer also provides prioritized schemes for isochronous and asynchronous data transfer. WiMedia UWB is designed for the personal computer, consumer electronics, mobile devices, and automotive market segments [20].

UWB Forum
DS-UWB uses a combination of a single-carrier spread-spectrum design and wide coherent bandwidth. DS-UWB uses the same spectrum as MB-OFDM (between 3.1 and 10.6 GHz) to transmit data rates up to 500 Mbps within a short range.

8.7   Other WLAN and WPAN Technologies

Besides the above-mentioned WLAN and WPAN solutions, other short-range wireless technologies are on the horizon. In this section, we briefly list some of them.

- *IrDA:* The Infrared Data Association (IrDA) was launched in 1994 to develop specifications for infrared wireless communication (as a cable replacement).

However, as the technology matured, it went far beyond being just a cable replacement tool and more toward enabling devices to exchange business cards, calendar items, and other object types. Today, IrDA is supported on many mobile phones, laptops, printers, personal digital assistants, and other products.

- *Radio-frequency identification (RFID):* RFID is a wireless technology that allows data to be collected via electronic tags. Unlike bar codes, the RFID tags can be read in the proximity of a transmitted radio signal. The tags or electronic labels can be passive or active. Passive tags do not require power, but can be tracked within only a few meters (for use in, say, grocery stores). In contrast, active tags do require power and can transmit RF signals up to hundreds of feet (for tracking items at, for instance, a loading dock).

- *ZigBee:* Zigbee conforms to the IEEE 802.15.4 standard for low-data-rate networks. It supports a peak data rate of 250 Kbps, which is lower than WiFi and Bluetooth, but its battery could last for months or even years. Home automation, remote control, and device monitoring are some key applications of ZigBee.

8.8 WLAN Versus WPAN

WLAN is designed to aid existing wired LANs and to be used where it is difficult to install cables. WPANs are envisioned to enrich our personal experience by providing wireless connectivity to two or more daily personal-use devices. WLAN is best suited for laptops, whereas WPAN is good for cell phones and PDAs.

The coexistence of both Bluetooth and 802.11b, the most popular WLAN standard, is tedious because both operate in the same unlicensed ISM 2.4 band. They compete for the same physical space and hence degrade each other's performance. This problem will be resolved when WLANs move to the 5-GHz band in the near future.

Some of the key differences between WLAN and WPAN systems are as follows:

- *Coverage:* WLAN technologies provide a longer range than the WPAN technologies. Bluetooth has a cell radius of roughly 10m, whereas 802.11 can easily cover 100m.

- *Data rates:* The peak data rate of Bluetooth is 1 to 2 Mbps compared to the 54 Mbps supported by 802.11b.

- *Power consumption:* WPAN technologies look for low-cost, low-power solutions contrary to WLANs.

- *Life span:* WPAN networks cannot exist without a master. If the master does not participate, the network will not be able to exist. WLANs, on the other hand, exist independent of their constituent devices.

8.9   Summary

This chapter provided an overview of WLAN and WPAN technologies. We discussed the activities of IEEE and ETSI standards bodies for such technologies. It also highlighted the differences between WLANs and WPANs, looked briefly at UWB, and provided details about IEEE 802.11, HIPERLAN2, and Bluetooth technologies. We will provide details about WLAN and WPAN security and interworking with 3G cellular in subsequent chapters.

References

[1]   Ware, C., et al., "The Evolution of 5 GHz WLAN Toward Higher Throughputs," *IEEE Wireless Communications*, Vol. 10, No. 6, 2003, pp. 6–13.

[2]   Prasad, R., and L. Munoz, *WLANs and WPANs Toward 4G Wireless*, Norwood, MA: Artech House, 2003.

[3]   Smith, C. and J. Meyer, *3G Wireless with WiMax and WiFi: 802.16 and 802.11*, New York: McGraw-Hill, 2004.

[4]   Crow, B. P., et al., "IEEE 802.11 Wireless Local Area Networks," *IEEE Communications Magazine*, Vol. 35, No. 9, 1997, pp. 116–126.

[5]   Garg, V. K., *Wireless Network Evolution*, Upper Saddle River, NJ: Prentice Hall, 2002.

[6]   Moreira, A., C., "The Infrared Physical Layer of the IEEE 802.11 Standard for Wireless Local Area Networks," *IEEE Communications Magazine*, Vol. 36, No. 12, 1998, pp. 107–112.

[7]   "IEEE Information Technology—Telecommunications and Information Exchange Between Systems—Local and Metropolitan Area Networks—Specific Requirements—Part 11: Wireless LAN Medium Access Control (MAC) and Physical Layer (PHY) Specifications," IEEE Standard 802.11, New York: IEEE, 1997.

[8]   Gu, D., and J. Zhang, "QoS Enhancements in IEEE 802.11 Wireless Local Area Networks," *IEEE Wireless Communications*, Vol. 41, No. 6, 2003, pp. 120–124.

[9]   Xiao, Y., "QoS Guarantee and Provisioning at the Contention-Based Wireless MAC Layer in the IEEE 802.11e Wireless LANs," *IEEE Wireless Communications*, Vol. 13, No. 1, 2006, pp. 14–21.

[10] Torsner, J., et al., "HiperLAN2: Broadband Wireless Communications at 5 GHz," *IEEE Communications Magazine*, Vol. 40, No. 6, 2002, pp. 130–136.

[11] "Broadband Radio Access Networks (BRAN); HIPERLAN Type 2; Data Link Control (DLC) Layer; Part 1: Basic Data Transport Functions," ETSI TS 101 761-1 (V1.3.1), ETSI, December 2001.

[12] McGeehan, J., et al., "A Comparison of the HIPERLAN/2 and IEEE 802.11a Wireless LAN Standards," *IEEE Communications Magazine*, Vol. 40, No. 5, 2002, pp. 172–180.

[13] Henry, P. S., and Luo, H., "WiFi: What's Next?" *IEEE Communications Magazine*, Vol. 40, No. 12, 2002, pp. 66–72.

[14] Fei, L., "RF Challenges for 2.4 and 5 GHz WLAN Deployment and Design," *Proc. IEEE Wireless Communications and Networking Conf.*, Orlando, FL, March 17–21, 2002, pp. 783–788.

[15] Sairam, K., N. Gunasekaran, and S. Reddy, "Bluetooth in Wireless Communication," *IEEE Communications Magazine*, Vol. 40, No. 6, June 2002, pp. 90–96.

[16] Karaoguz, J., "High-Rate Wireless Personal Area Networks," *IEEE Communications Magazine*, Vol. 39, No. 12, 2001, pp. 96–102.

[17] Haartsen, J., "Radio Network Performance of Bluetooth," *IEEE Int. Conf. on Communications*, New Orleans, LA, June 18–22, 2000, pp. 1,563–1,567.

[18]   Bisdikian, C., "An Overview of the Bluetooth Wireless Technology," *IEEE Communications Magazine*, Vol. 39, No. 12, 2001, pp. 86–94.

[19]   Porcino, D., and W. Hirt, "Ultra-Wideband Radio Technology: Potential and Challenges Ahead," *IEEE Communications Magazine*, Vol. 41, No. 7, 2003, pp. 66–74.

[20]   ECMA-368, "High Rate Ultra Wideband PHY and MAC Standard," ECMA, December 2005.

CHAPTER 9

Wireless Technologies Interoperability

9.1 Introduction

We have described a number of wireless access technologies in the last few chapters. A number of these technologies do run over each other and compete for the same space. The 3G vision was supposed to provide a single standard that works exclusively across the globe but we ended up with multiple, competitive solutions that do not interoperate with each other. Thus, the challenge for these different wireless access technologies is the design of a smart mobility management system that allows global roaming and session mobility among them (at least in the short term).

The focus of this chapter is on the mobility management and interoperability aspects of the technologies. The concentration will be aimed at the packet-switched domain (IP-based) aspects, with not much attention paid to circuit-switched domain attributes since most of the newer technologies are IP based. We will also address the potential mechanisms and shortcomings of interoperability among WWAN, WMAN, WLAN, and WPAN technologies.

It is worthwhile to note that many issues remain to be resolved to make interoperability effective. This chapter merely depicts possible opportunities for interoperability; it does not by any means provide detailed solutions. The standards bodies on the other hand are actively working to remove these hurdles and to build a stronger mobility management system.

9.2 Mobility Management

Mobility management covers a lot of ground and it has been exhaustively discussed by the wireless telecommunications industry at large. Several pieces are attached to MM, including the following:

- *Personal mobility:* This is the ability of the user to access his or her personal services (e.g., calendar, bookmarks) while away from the home network.
- *Session mobility:* This is about maintaining sessions when moving from one network to another (e.g., from 1xRTT to EV-DO, UMTS to WLAN).
- *Network and terminal mobility:* This is the ability of both the network and the terminal to support roaming.

9.2.1   IP Mobility

One key aspect associated with MM is IP mobility, which is in use in 2.5G/3G cellular and WLAN networks. IP is a connectionless protocol, meaning there are no distinctive paths between the transmitting and receiving ends [1]. IP works at both layer 3 (network) and layer 4 (transport) of the OSI model, although most of its benefits are associated layer 3 IP routing.

There are two versions of IP, IPv4 and IPv6, with IPv4 being the dominant network layer protocol on the Internet. The 3GPP2 specifications cover both versions of IP, whereas the 3GPP IMS is based around IPv6. Note that there are significant differences between the two, some of which are as follows:

- *IPv4:* IPv4 is a best effort protocol, that is, it does not guarantee delivery and does not make any guarantees on the correctness of the data. Thus, it may result in duplicated packets or out-of-order packets. IPv4 provides a limited number of 32-bit addresses, which are close to being exhausted. IPv4 encryption methods are proprietary and no standard encryption methods exist. Another challenge is the delivery of real-time multimedia services since a number of QoS methods have been used with IPv4. This means that all of the IPv4 QoS-compliant devices are not compatible with each other.

- *IPv6:* IPv6 addresses all of the shortcomings of IPv4. It uses 128-bit source and destination addresses to overcome the shortage of IP addresses. The security mechanism, IPSec, is available in some implementations of IPv4, but it is completely integrated into IPv6. IPv6 also supports a standardized QoS mechanism. Mobile IPv6 provides faster handover and better router optimization and mobility compared to mobile IPv4. Thus, IPv6 will be a better choice for 3G/4G and IMS-enabled systems.

Two types of IP mobility are defined, mobile IP (MIP) and simple IP (SIP). MIP is necessary to maintain sessions in macromobility environments. In MIP, both intra-PDSN and inter-PDSN address mobility are supported. On the other hand, in SIP only intra-PDSN address mobility is supported. The SIP service will drop the session in case of mobility, and the mobile node has to establish everything again. SIP is most appropriate with very limited mobility and where layer 2 mobility mechanisms satisfy mobility needs. The focus will be on MIP because it will be necessary for roaming among heterogeneous wireless access technologies. MIP has four main components, which are part of 3GPP- and 3GPP2-based networks but have different names:

- *Home agent (HA):* A server (stand-alone router) in the subscriber's home network. The HA is responsible for keeping track of where the subscriber is located at all times.

- *Foreign agent (FA):* Responsible for providing a temporary address (care-of address) to the visiting subscriber and for routing his or her packets to the new location. The FA is a router in the visited network. Its functionality is provided by PDSNs in the 3GPP2 architecture and by GGSNs in 3GPP-based networks.

- *Mobile node (MN):* Used to describe the end-user device. The device could be a data card, handheld device, laptop, and so forth.
- *Correspondent node:* Used for a device that is sending/receiving packets to/from the mobile node. A correspondent node could be an e-mail server, Web site, or even another MN.

9.3   3GPP2-Based Mobility Management

As stated earlier, the CDMA2000 (1xRTT and EV-DO) systems incorporate MIP (layer 3) to achieve mobility management. The MIP is used to provide seamless connections when roaming into a foreign network (WLAN or UMTS) and during inter-PDSN (FA) handoffs.

Location update, one of the functions of MM, is provided when the MN moves into a new PCF or new PDSN coverage area [2]. The PCFs are usually part of the BSCs, and a single PDSN usually supports more than one BSC. When the MN registers with the new BSC but within the same FA, the BSC establishes an A8/A9 connection to the new PCF. During inter-PSDN movement, the MN performs MIP registration with the HA as well as the BSC registration. To provide session mobility, session pipes are established between the MN and PDSN through a PPP connection. The PPP is only reestablished when the MN is switched to a different FA.

Because CDMA2000 1X systems are backward compatible with CDMAOne systems, there is no issue of handoffs between the two technologies. Even though 1xRTT and 1xEV-DO are not backward compatible with each other, data handoff is still possible with dual-mode handsets and data connection cards.

9.4   3GPP-Based Mobility Management

3GPP networks achieve MM a bit differently than 3GPP2 networks, but the concept of a HA and FA still exist. These networks (GPRS/EDGE and UMTS) use the combination of a proprietary protocol, GTP, and link layer (layer 2) management to offer IP mobility. GTP is used between UTRAN and SGSN, SGSN and GGSN, and between GGSNs. GTP adds an extra layer of overhead to the data, but does provide security and management of the packets both within the network and between the networks. The link layer management function is used to manage tunnels between SGSNs and GGSNs and provides a high degree of IP mobility as long as the user remains on the GPRS/UMTS network.

In UMTS, the base stations covered by an SGSN are divided into several routing areas (RAs). When the MN roams into a new RA, location update is performed and the RA is tracked by the SGSN. As far as session mobility is concerned in UMTS, a PDP context represents a session pipe between the MN and the GGSN. PDP consists of a PDCP tunnel between the MN and the UTRAN, and GTP tunnels between the UTRAN and the GGSN and between the SGSN and the GGSN. PDP supports delivery of packets between the MN and the core network.

9.4.1 UMTS: GSM Handover

In this section, we continue with our discussion of the UMTS to GSM handover mechanism, which was initiated in Chapter 4. First the users will need dual-mode handsets (UMTS/GSM) to provide such handovers from UMTS to GSM and vice versa. Second, the power consumption of such devices needs to be minimized. Specifically, when one of the modes is active (connected), the power of the other mode should be switched off.

Figure 9.1 presents a high-level pictorial of the UMTS/GSM intersystem hard handover process in the circuit domain. Like other handovers, this handoff is also initiated by requiring measurements of the parameters of the origin cell and the destination cell using the compressed mode. When the SRNC realizes that the target cell for the handover is not part of UMTS RAN, it sends the message "relocation required" to the MSC/VLR, indicating that the relocation should be performed in the GSM network. The MSC/VLR searches the BSS, maintaining the GSM cell to which the UE should be handed over, and sends this "handover required" message to the BSS. This results in the reservation of resources in the GSM MSC, which then sends an acknowledgment for the same back to UTRAN. A GSM handover command is then initiated by UTRAN, effectively transferring the call to the GSM BSS [3].

As stated in Chapter 4, the compressed mode consists of interrupting the transmission and/or reception in order to allow the mobile to change the frequency during blanking. Depending on the UE implementation and the measured frequency, the CM can be activated either in the downlink or uplink or both. The interruption of transmission/reception is called a *transmission gap,* and it lasts from 3 to 14 slots. The transmission gap distribution is defined by two transmission gap patterns that are repeated successively a certain number of times [4, 5]. This succession of the two repeated patterns is called the transmission gap pattern sequence (TGPS). At the end of the TGPS, the CM is deactivated, and the UE switches back to its previous carrier's uplink and downlink.

To avoid loss of data, the data transmission time is reduced either by reducing the spreading factor, puncturing, which can only apply in downlink, or lowering the data rate. The common technique that is used to avoid data loss is reduction of the spreading factor, which allows data transmission but at the cost of higher power.

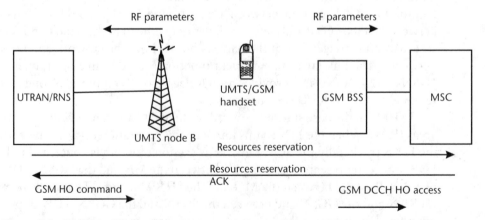

Figure 9.1 UMTS-to-GSM handover.

The requirement for increased power has a negative impact on the coverage and capacity of the system.

The duration required to perform measurements on GSM cells depends on the CM pattern and also on the number of cells in the neighbor cells list. If the gaps are frequent, this duration will be shorter but it will degrade system performance, and if the gaps are longer, it could result in delay of the handover execution and loss of connection. Thus, finding a compromise is important to avoid a sizable reduction in system performance. It is also important to note that the sooner the measurements on GSM are triggered, the lower the call drop rate. Some additional results on QoS, call drop rate, receiver sensitivity, and so forth can be found in [5].

In the packet-switched domain, the handovers are not supported in real time due to lack of support in GSM/GPRS. However, the connection is kept and packets are forwarded from Iu PS to GPRS Gb or vice versa under control of the serving SGSN. In addition, GPRS only supports a 171.2-Kbps peak data rate when all eight timeslots are available for data, which is not the case in most circumstances. All of these factors result in data throughput degradation when data calls are transferred to the GPRS network.

The handovers from GSM to UMTS do require a similar process, but many of the current GSM handsets cannot work in the UMTS network. Modifications also need to be made in the GSM BSS and MSC to support handovers to UMTS.

9.5   CDMA2000 and UMTS Mobility Management

The deployments of WCDMA are finally taking place, and it is time for both CDMA2000 and UMTS operators to diligently think about interoperability between the two. Though the increased usage of data services and enablement of VoIP will likely switch most of the attention to IP services in the coming years, interworking has to be addressed at both the circuit-switched and packet-switched levels. Thus, we briefly consider here the interoperability aspects of both domains for the two technologies.

There are three types of mobility in CDMA2000 and UMTS systems: radio mobility, core network mobility, and IP mobility. Radio mobility supports handoff (i.e., radio link switching) of a mobile user during conversation. CN mobility (or link layer mobility) provides tunnel-related management for packet rerouting in the CN due to user movement. The IP mobility mechanism allows the mobile user to change its access point of IP connectivity without losing ongoing sessions.

The overall architecture of MM is similar in both UMTS and CDMA2000. However, the protocols exercised on the network architectures are different. For example, CDMA2000 uses IETF protocols (e.g., MIP and PPP) to support MM mechanisms, and it is connected to an ANSI-41-based network protocol. In contrast, the UMTS protocols include SS7-based MAP and IP-based GTP. Moreover, one of the key 3G goals is to provide independence between the radio and core networks. UMTS provides a clear demarcation between the RAN and CN, whereas this goal is only partially achieved in CDMA2000 where the MSC is still involved in radio resource allocation for a packet session [2, 6].

To achieve interoperability between UMTS and CDMA2000, the interworking issues at the radio, CN, and IP levels must be addressed. For radio level interworking, a dual-mode terminal that contains both UMTS and CDMA2000 radio interfaces and the capability to perform real-time handoff among these two systems is required. For CN interworking, the MM and PDP contexts have to be migrated between UMTS and CDMA2000 networks. For IP-level interworking, roaming between the two networks must be supported.

A simple (short-term) approach to address some of these shortcomings is to introduce a gateway and HA between the two networks for the CS and PS domains, respectively. Figure 9.2 depicts the gateway/HA solution that can be considered as the simplest way of implementing interworking between two dissimilar technologies, without hardly any modifications to the existing CDMA2000 systems and the new UMTS systems. In this figure, the UMTS network connects to the IP network through the GGSN that acts as a MIP FA. The MS first registers with the HA via the GGSN (MIP FA). The HA maintains the MS's location information (e.g., the GGSN [MIP FA] address) and tunnels the IP datagrams to the MS. Both the UMTS and CDMA2000 networks connect to the packet data networks through the MIP HA. Therefore, when a data call is initiated, it is first delivered to the MIP HA. Then the MIP HA forwards these packets to the MS through the MIP FA (GGSN or PDSN). The gateway solution has its shortcomings, but it will remove some of the difficulties of the mobile workforce until the arrival of 4G, as we will discuss in a later chapter.

9.6 3GPP and TD-SDMA Interworking

From the very start, TD-SCDMA was designed to allow easy integration with both GSM and WCDMA networks. If the TD-SCDMA radio network is connected to a GSM core, it is called TSM, and if it is hooked up to a UMTS core it is termed TDD_{LCR}, as stated in Chapter 5.

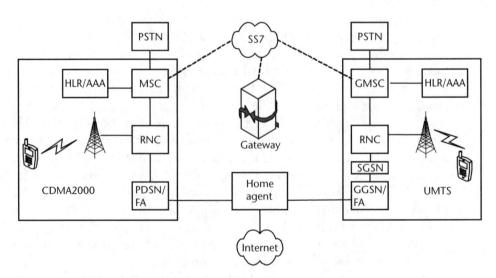

Figure 9.2 UMTS and CDMA2000 interworking.

The TD-SCDMA base stations can be connected to the GSM BSCs, which are upgraded to support them by means of the software TGCF. The new BSCs are connected to the GSM core network by the existing A and Gb interfaces. These upgrades offer seamless interworking and support handovers between TD-SCDMA and GSM/GPRS radio systems.

From the UMTS perspective, the TD-SCDMA base transceiver stations are connected to the TDD_{LCR}, which enables RNCs. These RNCs support the A and Gb interfaces to connect to the MSC and GGSN/SGSN and the Iu UMTS interfaces (Iu-cs and Iu-ps) to connect to the UMTS core network.

9.7   3GPP and WLAN Interoperability

In this section we consider two cases of interoperability: GSM/GPRS and WLAN, and UMTS and WLAN.

9.7.1   GSM/GPRS and WLAN Interoperability

Most of the interoperability work has been conducted between GSM/GPRS and WLAN networks. In fact, a consortium of companies have developed a set of open specifications for the adoption of the same, that is, unlicensed mobile access (UMA) technology. This UMA effort was initiated in January 2004, and a completed set of specifications was issued by late 2004. UMA provides access to GSM and GPRS mobile services over unlicensed spectrum technologies that include Bluetooth and WiFi via dual-mode handsets. In 2005, the specifications were formally approved by 3GPP and were included into 3GPP Release 6.

9.7.1.1   GAN Architecture

Figure 9.3 shows the generic access network (GAN), which consists of one or more APs and one or more generic access network controllers (GANCs), interconnected

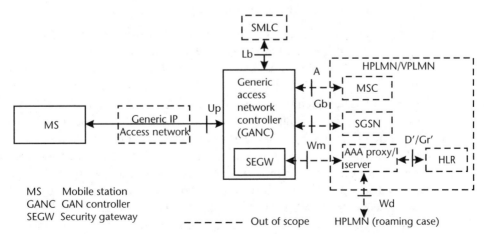

Figure 9.3   GAN functional architecture [7].

through a broadband IP network. Within 3GPP, the UMA technology specification is known as GAN, not UMAN.

The GAN architecture requires a dual-mode handset (GSM/GPRS plus UMA-enabled) and a new element called GANC, which appears to the core network as a GERAN BSS. GANC includes a security gateway (SEGW) that terminates secure remote access tunnels from the MS, providing mutual authentication, encryption, and data integrity for signaling and voice and data traffic. A broadband IP network provides connectivity between the AP and the GANC. The IP transport connection extends all the way from the GANC to the MS, through an AP.

A new single interface, the Up interface, is defined between the GANC and the MS. The A interface and Gb interface for coexistence with GERAN and interconnection with the GSM core network are still used. The A and Gb interfaces support CS and PS services, respectively. The Wm interface is used over the AAA server, which authenticates the MS when it sets up a secure tunnel. Also, in this architecture, functionalities such as call processing and user services are provided by the network elements in the core network, namely, the MSC/VLR and the SGSN/ GGSN [7].

9.7.1.2   How GAN Technology Works

When the dual-mode handset finds a GAN network, it contacts the GANC for authentication over the broadband IP network. Upon authentication the MS is allowed to access GSM voice and GPRS data services. Also when the MS is approved, the subscriber's current location information stored in the core network is updated, and from that point on all services are routed to the handset via GAN rather than RAN. This architecture supports seamless handovers from GSM/GPRS to GAN. These cross-technology handovers are called *vertical handovers* compared to *horizontal handovers*, which occur between the network elements of a homogeneous wireless system. In the case of roaming, when a subscriber moves outside the range of an unlicensed wireless network to which he or she is connected, the GANC and handset facilitate roaming to the licensed outdoor network.

9.7.2   UMTS and WLAN Interoperability

The interoperability between UMTS and WLAN can be addressed with mobile IP and dual-mode MSs, as shown in Figure 9.4. Mobile IP is used to reshape connections when an MS roams from one network to another [8]. The MS is identified by a care-of address (COA) when it is outside its home network. The MS registers its COA with the HA, which resides in the home network of the MS. The HA is responsible for intercepting datagrams addressed to the MS's home address as well as encapsulating them with the associated COA. The datagrams to an MS are always routed through the HA, whereas datagrams from the MS can be tunneled via HA.

Two types of vertical handovers can occur between the technologies, namely, upward vertical handover and downward vertical handover. An upward vertical handover occurs from WLAN to UMTS, while a downward vertical handover takes place from UMTS to WLAN. In the case of an upward vertical handover, the MS finds that it cannot locate beacons from the serving WLAN and instructs the WLAN

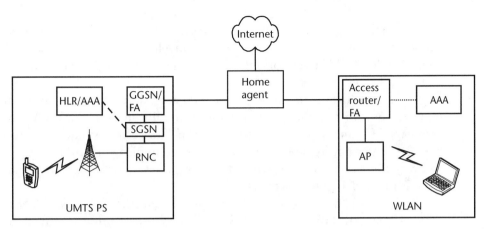

Figure 9.4 WLAN and UMTS interworking.

to stop forwarding packets. The MS routes this instruction via a MIP registration procedure through the UMTS CN and hands over control to UMTS network. When it is connected to the UMTS network, the MS listens to the lower layer WLAN access point. If several beacons are received successfully, it will switch to the 802.11 network via the MIP registration process. Thus, the vertical handover decisions are based on the presence or absence of beacon packets.

The 3GPP has defined two procedures to interwork with WLAN: WLAN direct IP access and WLAN 3GPP IP. These 3GPP-WLAN interworking architectures are expected to be independent of the underlying WLAN radio technology. In addition, the support of QoS is optional in the 3GPP-WLAN interworking architectures. The details can be found in [9].

- *WLAN direct IP access:* In this procedure, the WLAN radio access and authentication and authorization, which provide access to the WLAN and the locally connected IP network (e.g., Internet), will be authenticated and authorized through the 3GPP System.
- *WLAN 3GPP IP access:* This procedure allows WLAN terminals to establish connectivity with external IP networks, such as 3G operator networks, corporate intranets, or the Internet via the 3GPP system.

9.8 CDMA2000 and WLAN Interoperability

In most cases WLAN and CDMA2000 radio technologies provide complementary environments for voice and packet data services. A CDMA2000 system provides broad coverage and better authentication and accounting mechanisms. A WLAN system provides localized coverage and thus complements the coverage of a CDMA2000 system. Because WLAN networks typically cover small geographic areas, the WLAN providers can take advantage of certain capabilities provided by CDMA2000 systems. Conversely, the CDMA2000 system operator may offload some data traffic to WLAN networks when the users with dual-mode devices are in the WLAN coverage area. Keeping these views in mind, the 3GPP2 has recently completed the stage 1 requirements for CDMA2000 and WLAN interworking [10]. These requirements define a logical model (Figure 9.5) that shows the CDMA2000-WLAN interworking relationships, which may be either direct or indirect (i.e., through a broker system[1]).

Figure 9.6 defines a more detailed (nonstandardized) integration approach between the two technologies that requires modification of the WLAN portal (or gateway), HA, and new client software for the devices. To support intertechnology handoffs, the gateway, HA, and devices have to implement MIP functionality. The gateway connects to the HA, which maintains the station location information and tunnels the IP datagrams to the station.

Both WLAN and CDMA2000 networks connect to the packet data networks through the MIP HA. Therefore, when a data call is originated from a WLAN

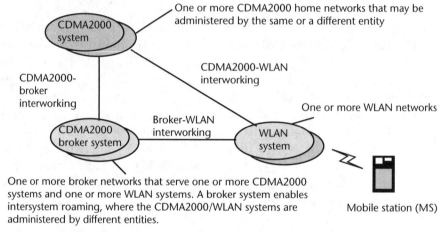

Figure 9.5   CDMA2000 and WLAN interworking model [10].

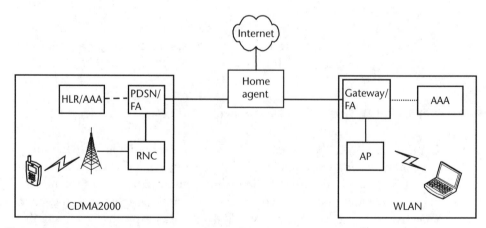

Figure 9.6   WLAN and CDMA2000 interoperability.

---

1.   A *broker system* is a collection of intermediaries that facilitates WLAN/CDMA2000 interworking, where CDMA2000 and WLAN systems are administered by different entities.

station, it is first delivered to the MIP HA via gateway. Then the MIP HA forwards these packets to the CDMA2000 handset through the PDSN. In this setup, the gateway also has to support AAA services to interwork with CDMA2000 AAA servers. This will enable the 3G service provider to collect the WLAN accounting records for generating a unified bill. Also, the use of compatible AAA services on the two networks would allow the WLAN gateway to dynamically obtain service policies on a per-user basis from the 3G AAA servers, which then can be enforced to the WLAN network [11].

9.9   Overall Picture of Wireless Technology Interoperability

In the earlier sections, we addressed the interoperability aspects between WWAN and WWAN and between WLAN and WWAN technologies. Though less studied, interoperability between WMAN (IEEE 802.16d/e and IEEE 802.20) and WWAN (UMTS/CDMA2000) has to be considered in the near future. Similarly, if WMANs are deployed in the coming years they also need to interoperate with 802.11-based technologies. The coexistence aspects of WLANs and WPANs as discussed in the previous chapter have to be further studied to understand their merits. The handovers among these technologies will most likely be based on MIP. Figure 9.7 shows an overall interoperability picture for the different access technologies.

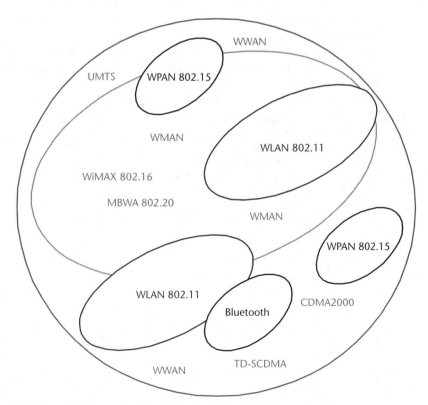

Figure 9.7   Overall picture of system interoperability of wireless technology.

9.10 Summary

In this chapter, we defined the interoperability aspects of the different wireless technologies. The focus was primarily on the interoperability aspects of WLAN and WWAN technologies. We also discussed the mobility management features of 3GPP- and 3GPP2-based systems. We believe that MIP is a must-have feature if the operators want handshakes with other operators to generate more profits for their own organizations. At the same, all of the interoperability methods have both strengths and weaknesses, as discussed in this chapter, that have to be resolved in the coming years. Finally, work has to be conducted to look into the handshakes among 802.11- and 802.16d/e-based technologies.

References

[1]   Thelander, M., "IPv6—The Path Less Traveled," *Signals Ahead*, Vol. 2, No. 15, August 1, 2005.

[2]   Agrawal, P., et al., "Mobility and Session Management: UMTS vs. CDMA2000," *IEEE Wireless Communications*, Vol. 11, No. 4, August 2004, pp. 30–43.

[3]   Willie, W. L., *Broadband Wireless Mobile*, New York: John Wiley & Sons, 2002.

[4]   "Physical Layer—Measurements (FDD)," Technical Specification (Release 1999), Technical Specification Group Radio Access Network, 3GPP TS 25.215 (V3.11.0), 3GPP, September 2002.

[5]   Lugara, D., et al., "UMTS to GSM Handover Based on Compressed Mode Technique," *IEEE Conf. on Communications*, Paris, France, June 20–24, 2004, pp. 3,051–3,055.

[6]   Lee, S., et al., "Interoperability Between UMTS and CDMA2000 Networks," *IEEE Wireless Communications*, Vol. 10, No. 1, February 2003, pp. 22–28.

[7]   "Generic Access to the A/Gb Interface; Stage 2," Technical Specification (Release 6), Technical Specification Group GSM/EDGE Radio Access Network, 3GPP TS 43.318 (V2.0.0), 3GPP, January 2005.

[8]   Prasad, R., and L. Munoz, *WLANs and WPANs Toward 4G Wireless*, Norwood, MA: Artech House, 2003.

[9]   "3GPP System to Wireless Local Area Network (WLAN) Interworking; System Description," Technical Specification (Release 7), Technical Specification Group Services and Systems Aspects, 3GPP TS 23.234 (V7.2.0), 3GPP, June 2006.

[10] "CDMA2000—WLAN Interworking; Stage 1 Requirements," 3GPP2 S.R0087-A (V1.0), 3GPP2, February 2006.

[11] Miller, S., et al., "Design and Implementation of a WLAN/CDMA2000 Interworking Architecture," *IEEE Communications Magazine*, November 2003, pp. 90–100.

CHAPTER 10

Wireless Technologies: From Inception to Deployment

10.1 Introduction

In the previous chapters, different forms of wireless technologies were discussed. The concentration was mainly on the technologies themselves and not on the process that the technologies are tested against before they are finally delivered to the end customers. Therefore, the focus in this chapter is on the different steps that are required before a wide-area technology solution is delivered to an end customer.

We will use CDMA2000 1X (3G) as an example to describe the process specifically from an operator's perspective. The implementations of WLANs or WPANs are not considered, and the discussion is mainly tied to the deployment of a 3G WWAN technology. The process is still quite lengthy, especially if the operator has a huge network and multiple original equipment manufacturers (OEMs) and infrastructure suppliers, which require the involvement of many internal groups and a large workforce. We consider here all of the key steps that are required to launch a 3G system. The multiple steps involved from start to finish include the following:

- Standardization of the technology;
- Research and development phase (early technology trial);
- Field (market) trial;
- Deployment;
- Operations.

We attempt to unfold many aspects of these steps as we describe the overall process. So let's start from the very first step, *standardization*.

10.2 Standardization of the Technology

Standardization in many cases is the first step before a new technology can be implemented on a large scale. In our case, the standardization work for CDMA2000 1X (short form: 3G1x) started in the mid-1990s. The OEMs, device manufacturers, chipset suppliers, even the operators—all of them submitted their proposals to 3GPP2. Because the changes represented a major transition from CDMAOne (2G), the agreement on a single collective proposal took a long time. The suppliers spent a considerable amount of time supporting their claims with simulations. In 1999, the 3GPP2 standard body finalized the 3G1x standard and called it IS-2000 Release 0.

The IS-2000 standard describes the physical, MAC, and upper layers and analog operation. Although there are other standards (such as the Interoperability Specification for CDMA 2000 Access Network Interfaces [3G-IOS] and IS-835) in the process, the key one is IS-2000, which started all of the activity.

10.2.1 Key Aspects of 3G1x

The key improvements of 3G1x over 2G are as follows:

- The CS voice capacity is doubled; that is, the number of users in a cell is doubled.
- A data capacity gain is realized due to migration from CS data to PS data.
- The symmetrical service supports 153-Kbps peak data rates from tower to mobile and mobile to tower.
- Paging functions are improved.
- Mobile battery life is improved.

10.3 Research and Development Phase

Most of the operators around the world do not actually develop products; they instead work with rest of the industry to develop the products for their end customers. The operators mainly own the frequency spectrum and the network (3G1x components) that provide voice, video, and data services. Some operators conduct proof of concept (POC) or early technology trials to validate the technology and the standard. The validation of 3G1x technology was conducted across the globe by the telecommunications industry. The early technology trials were conducted to understand and validate the enhancements of 3G1x.

These tests are mainly done in labs, and a smaller over-the-air (OTA) test bed (noncommercial network) can also be used by for this purpose. The OTA test bed, as shown in Figure 10.1, is usually comprised of a handful of cell sites. The testing is conducted by R&D teams and is not extensive but provides an early look into the technology. It helps the operator and its OEMs to improve the technology before the next phase of the field trial. Note that the testing at this point does not have to look into the interoperability aspects among suppliers.

Some key aspects of the 3G1x air interface that can be tested in this phase are as follows:

- Analysis of voice quality, capacity, and coverage;
- Validation of physical layer data rates and data throughputs;
- Validation of application layer data rates and data throughputs as perceived by the user;
- Analysis of mixed voice and data coverage and capacity;
- Measurement of battery life of handsets and talk time of different handsets;

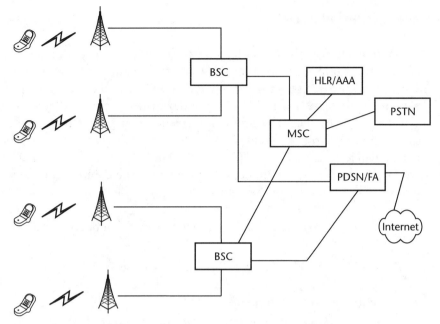

Figure 10.1   POC OTA test bed configuration.

- Performance of the RF scheduler, which resides in the BS (every OEM has a different way of implementing this feature so it will prudent to test across all suppliers);
- Handoffs between 2G and 3G;
- Backward compatibility with 2G.

10.3.1   Process

The operators usually send a set of requirements to their suppliers that they would like to validate. The technology evaluation document lists the high-level trial objectives, technical and functional requirements, test cases for technology validation, roles and responsibilities of the R&D team and suppliers, architecture of the test bed, software and hardware requirements, and so forth. After the agreement on the set of objectives and requirements, the next step is to develop the test plan.

The test plan provides detailed information on the test cases that were initially defined in the technology evaluation document. It describes the entrance and exit criteria, test bed architecture, procedures for all of the test cases, equipment, software loads, pass and fail criteria for test runs, and so forth. The operators and the suppliers have to sign off on the joint test plan before they can move forward. The suppliers work with the R&D team to install and perform hardware testing of the equipment in the lab and also two to three OTA test bed noncommercial cell sites (if any). Test execution usually takes 4 to 6 weeks and is completed with the delivery of a detailed test report and presentation to the executives. After the completion of this phase, the network development team conducts field testing across all infrastructure suppliers for more rigorous validation.

10.4   Field (Market) Trial

After completion of the technology trials, operators can conduct field trials in market(s). For example, Figure 10.2 shows that a U.S. operator has four infrastructure suppliers to provide services nationwide. It will be prudent for the operator's R&D team to validate 3G1x across the four and provide specific feedback to each of them. This testing is conducted by a network development team across the four infrastructure suppliers in four different markets. Suppose, infrastructure supplier A is in San Diego, B in Kansas City, C in Toledo, and D in Jersey City. The testing across suppliers can be conducted simultaneously to expedite the deployment process.

Some key elements of the field trials are as follows:

- Acquire a separate carrier (frequency) to avoid impacts on the production network and customers.
- Select a cluster of cell sites, maybe 19 cells for the trial.
- Conduct site surveys to make sure that existing sites have enough room for additional equipment and so forth.
- Gather all necessary documentation from the specific suppliers well before the commencement of testing.
- Select a couple of handsets from different handset manufacturers to test against.
- Certify the devices that will be used by the end customers.
- Conduct alarm testing, operational measurement testing, billing testing, provisioning testing, back-office testing, and testing of installation in addition to the capacity and coverage tests that were conducted in the R&D phase.

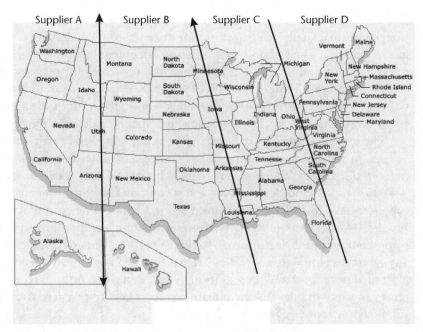

Figure 10.2   Example of operator coverage across the United States.

- Perform end-to-end performance testing.
- Conduct intervendor handoffs (e.g., a user moving from Kansas City to Pittsburgh).
- Capture all of the trial results and document their results.

The IT organizations of the operators usually develop software that can be used to provide user security, provisioning and billing, and so forth. Sometimes operators also outsource this software development to a third party. After development and high-level testing, these features are implemented in the network and then exhaustively tested by the network team in collaboration with the IT organization.

10.5 Deployment Phase

This is the longest phase of the process, and it involves almost all departments of a company. The rollout of a 3G1x technology by an operator could take a year or so if it has to be deployed nationwide. Several pieces of the puzzle need to be solved and clarified before it can be handed over to the operations team for day-to-day management. The details of two key aspects of deployment, network planning and network management, which are handled by the engineering and operations teams, are discussed in the following subsections. We will also cover some of the other key pieces that are necessary to launch the network.

- *Program management:* The program management (PM) organization has the responsibility of managing the entire deployment process from start to end. It coordinates testing, ensures equipment delivery, organizes meetings, ensures the smooth flow of information from one organization to another, and finally delivers the role to operations.
- *Spectrum management:* To overlay the 3G1x network over 2G, the operator has to ensure that it has the entire necessary frequency spectrum in all markets. If it requires some additional spectrum in any area, then it should work with the regulatory bodies ahead of time to overcome the shortage. Similarly, the department should have a 3- to 5-year spectrum forecast to address any shortages of carriers if the demand rises.
- *Finance and SCM:* The finance and supply chain management (SCM) teams will be required to work with the OEMs, test equipment, device suppliers, contractors (if any) to figure out the overall deployment cost for the entire network. This deployment cost either has to be financed by borrowing money from potential investors, or the operator makes the investment himself. The negotiations with the suppliers take a considerable amount of time; efforts should be made to start the process early. Further, operators have to develop strong relationships with contractors so that network development can be completed in a timely manner. The longer it takes to construct and deploy the network, the higher the cost.
- *Marketing and sales:* The marketing and sales team needs to be educated about the product and technology. The sales team would be in charge of generating leads, especially corporate clients. This process should be started well

in advance, possibly three to four months before the actual "live" date. Besides the brochures and product literature, it would be useful to have a demo setup for the new product so potential clients can experience the technology firsthand.

The marketing team needs to come up with the subscriber forecast for the next 5 years. Before the forecast can be created, this team would need to define the target customers as well as target regions. Once target regions and customers have been defined, the new technology would need to be marketed through various advertising channels. This includes TV ad campaigns with multiple advertisements, promotional in-store campaigns, and partnerships with mass retailers (e.g., BestBuy, RadioShack). In recent times, word-of-mouth marketing has played a crucial role in rapid adoption of the technology, especially if a buzz can be created regarding the product. Therefore, this mode of advertising should be considered too. In the end, the goal is to create a distinct brand identity and ensure a successful launch of the service (3G1x).

- *Product development:* The product development (PD) team is responsible for the selection of device suppliers and devices. It is also responsible for supplying them with proper documentation and design requirements. The 3G1x devices mainly include cell phones and air cards for the customers. The network engineering team evaluates and thoroughly tests these products to ensure that high-quality products are delivered to customers. It is essential that strong, long-term relationships be developed with core device suppliers. This will ensure that the supplier develops high-quality products.

- *Network engineering:* This organization is responsible for engineering of the network (in our case, 3G1x), specifications for the enhancement of existing cell sites, supplier management, planning of coverage and capacity, and so forth. It also has the responsibility of delivering leading-edge features and products in a timely and cost-effective fashion. The handset team within network engineering is responsible for validation and testing of the devices that will be used in the network.

- *Field and RF engineering:* Field engineering installs wireless infrastructure equipment, acquires new cell sites, maintains and repairs equipment, and tracks all network assets, including spares for all the network elements. It maintains switch sites, cells and carriers, the wireless infrastructure, and backhaul (T1 lines). In addition, it is also responsible for designing and delivering built-in coverage for customers at their business locations and in public spaces, such as airports, convention centers, hotels, and shopping malls. RF engineering within field engineering has the responsibility for capacity management, RF optimization, drive testing, test equipment selection, site zoning, leasing, and so forth.

- *Network operations:* Network operations is responsible for surveillance, remote management, and maintenance of the wireless (3G1x) networks, including radio stations, cell site towers, and switches. This organization also manages transport platforms and data centers and the service assurance functions in case of network failure. It is also responsible for deployment, removal, testing, and acceptance of all new network equipment.

• *Information technology organization:* The information technology (IT) team within an operator is responsible for developing, testing, and maintaining the accounting and billing software. In addition to maintaining the company's intranet and Internet Web sites, it also develops tools for provisioning devices and network elements.

10.5.1   Network Planning

Network planning consists of four major components: radio network planning, network dimensioning, network optimization, and future expansion planning. Preliminary radio planning consists of cell site allocation, traffic planning, and extensive radio network planning. In network dimensioning, the required number of channel elements in the base station, the capacities of transmission lines, and the number of BSCs, MSCs, PDSNs, routers, and so forth are calculated. The network optimization is used to deliver the best quality of service possible to the customers. The last step in network planning is the forecast for the future expansion of the network for additional subscribers, newer features, and better technologies [1, 2].

10.5.1.1   Preliminary Radio Network Planning

The prerequisite for good radio network planning (RNP) is the excellent know-how of the RF environment. The radio coverage is usually simulated before investments are made in procurement of cell site locations and equipment. Because the radio environment is highly unpredictable, detailed measurements are needed for each individual cell. In addition, coverage, capacity objectives, and a network planning strategy need to be defined. The RNP can be divided into the steps discussed in the following subsections.

Cell Site Allocation
Cell sites must be designed and selected carefully so that they can conform to the government guidelines. The radio coverage must be such that smooth handovers can take place between cell sites to cover the entire area. Radio planning software and drive testing can be used during the cell site placement process.

The first step is to plan a theoretical cell site design using radio planning software that can simulate the coverage properties of the cell sites. The next step is to find actual locations based on the simulations on which to build real sites. Sometimes the theoretical cell sites may be located in areas where real cell sites cannot be constructed. During this physical inspection, other items such as terrain, existing structures, power, and zoning factors also need to be considered.

After the physical inspection the design may have to be retuned and, therefore, the initial cell site design is highly important. During the network design process, the country's radio regulatory policies and other restrictions such as maximum tower height and neighborhood laws must be strictly followed. Tower heights of 300, 150 to 200, and 100 feet are cost effective for rural, suburban, and urban areas, respectively.

Some of the challenging nontechnical tasks of deployment are site identification, acquisition, and construction. An unavailable cell site can toss months of planning down the drain. Therefore, the whole task of cell site allocation and commissioning should be carefully laid out to meet the launch date.

Traffic Planning

The wireless subscribers' signal strength can be measured based on the population of the given area, the population working in a given area, and vehicle traffic. Dense urban locations such as downtown areas and business parks have to be projected to have higher cell sites count than residential and rural areas. All of this ensures that the network that is about to get constructed will be able to provide services to all of its subscribers.

The next step is to estimate the cell capacity and cell coverage. The number of cells required to cover a certain geographical area is based on capacity and the link budget. A network can either be capacity or coverage limited. A capacity-limited network is one in which the maximum cell radius cannot support the total offered traffic. A coverage-limited network is one that has enough capacity in a cell to support all traffic, but the maximum cell range limits it.

Cell capacity can be estimated based on simulations or analytical formulas. A CDMA network's capacity is usually determined using the pole capacity. Pole capacity is defined as the theoretical maximum capacity. Another aspect that needs consideration is the load factor, which tells us how close to the maximum capacity the network can operate. The CDMA2000 systems usually operate at 50% to 60% of pole capacity.

The link budgets are used to calculate the coverage (range) for the downlink and uplink. Asymmetric traffic has to be considered in the link budget calculations. Normally, there is more downlink traffic than uplink traffic, particularly in the case of data traffic. To manage the different link budgets and thus different ranges, the downlink link budget is matched to the uplink link budget by adjusting the BS transmission power.

Extensive Radio Network Planning

Detailed network planning for CDMA2000 networks is conducted using software planning tools such as Decibel Planner. "Test drives" have to be performed to determine how the site will perform in the network by capturing the propagation characteristics of the site. These test drive data can be further compared with Decibel Planner's propagation results to accurately simulate the cell site's coverage. Link budgets also have to predict the range of the cells.

Detailed network planning consists of the following steps:

- *Characterization of the radio environment:* The radio environment has to be characterized in detail by the actual field measurements. Characteristics such as path loss, shadowing, and multipath need to be studied. In addition, an RF survey can be performed using spectrum analysis, which reveals competing cellular operators and intermodulation products.
- *Overhead channel planning:* The transmission of control channels reduces the overall capacity of the network. In CDMA2000 systems, roughly 20% to 25% of the total BTS power is allocated to the overhead channels (pilot, paging, and SYNC), while the remaining 75% to 80% is allocated to the traffic

channels. The power allocation to the pilot, paging, and SYNC channels is approximately 15%, 2% to 3%, and 5% to 7%, respectively.

• *PN code planning:* Unlike analog systems that use frequency and/or timeslots to differentiate the channels, 3G systems use different code sequences. This means that 3G channels require code planning instead of frequency planning. In 3G systems, PN code offsets are used to identify different sectors. Cell sites can share the same code value, but must be separated by enough distance to avoid interference. A mobile differentiates the pilots by their PN offset. Proper PN offset planning is needed to avoid an alias of PN sequences in a CDMA network.

• *Pilot pollution:* Pilot pollution occurs when there are multiple pilot signals, but none of them is dominant enough to allow the mobile station to start a call. This situation occurs when there is good mobile receiving power, poor E_c/I_o (energy/interference), and poor forward BER. The pilot pollution can be avoided by scaling the pilot powers, downtilting antennas, or increasing coverage of certain sectors or cells.

• *Soft handoff parameter planning:* Soft handoffs impact the required fade margin against shadowing and the number of users in soft handoff. It is the margin required to provide a specified coverage availability over the individual cells. It is one factor that influences the number of base stations that have to be deployed. Users in soft handoff require additional base station resources and backhaul connections. The goal is to minimize the fade margin and number of users in the soft handoff, while maintaining a satisfactory quality of service.

• *Hard handoffs:* There are a few various types of hard handoffs, including interfrequency, intervendor, and intertechnology. The main parameter during an interfrequency hard handoff is the frequency at which the base station commands the mobile station to measure neighboring carriers. The stronger the serving cell, the less frequently the mobile would check the neighboring channels. At the cell edge, the mobile measures the pilot channels more often in order to be prepared for the interfrequency hard handoff. The intervendor hard handoff occurs when the user moves from one vendor's MSC or PDSN to another vendor's MSC or PDSN within the network of a single operator. The intertechnology handoffs, such as from CDMAOne to CDMA2000 or vice versa, take place when the user leaves the area of one technology and enters the boundaries of the other. Because the same base stations can support both technologies, those are configured to support this handoff.

• *Indoor coverage planning:* Field measurements are required to characterize individual buildings because no general propagation model exists that can cover all different types of structures. The deployment of indoor cells in the same frequency as outdoor cells is a challenging task. More careful planning is needed when indoor cells are on the same frequency as outdoor cells. Planning of microcells and macrocells in the same frequency is possible with low antenna installations and antenna tilting.

• *2G and 3G coexistence:* The 2G and 3G networks that are deployed in the world are either in the same frequency band or in different bands. The CDMAOne and CDMA2000 systems can be deployed in the same or different frequency bands, whereas GSM and W-CDMA systems have to be deployed in different bands. The key aspects of intermodulation, guard bands, and handoffs have to considered when designing a mixed 2G/3G network.

- *Sectorization:* Normally the CDMA cells are made up of three sectors, which increases the capacity of the system. However, the number of soft handoffs also increases with sectorization, so it requires good planning for proper implementation. Several factors affect the gain, including interference, location of antennas, and coupling between antennas.

- *Handset verification:* Another important aspect is the validation of multiple devices during the optimization phase. Subscribers usually blame the network when they lose calls or when the quality is bad, but on some occasions the device itself is the main cause of poor service and not the network. Thus, the handset team must validate the performance of multiple types of phones (single and multiple handset suppliers). The CDMA2000 handsets normally have 200-mW RF output into a dipole antenna. A height of 1.5m above the ground, which is roughly that of a subscriber seated in a car, can be used when modeling propagation.

After all of these planning exercises have been carried out, a detailed coverage analysis can be carried out by evaluating frequency reuse coefficients. The frequency reuse coefficient is the ratio of in-cell to total interference and it is unique for each cell. These are then used again to predict the coverage within the different cells. Cells are designed with 10% to 15% overlapping coverage to account for cell breathing. An 8-dB fade margin often provides better than 90% area coverage.

10.5.1.2   Network Dimensioning

Another key aspect in the whole process is network dimensioning, which assists in BTS channel element planning, BSC planning (core network), back-office elements, backhaul (transmission) capacity, and coexistence with 2G as necessary.

- *BTS channel element planning:* The number of channel elements can be calculated based on voice and data traffic, overhead channels, and soft handoff factors. Softer handoff, on the other hand, does not require an additional channel element because the same channel element can handle signals from two or even three different sectors of the same base station.

- *BSC, core network, and back-office elements planning:* The number of BSCs, core network (MSC, PDSN, GGSN, SGSN, and so forth), and back-office (HLR, VLR, and so forth) elements is based on the traffic load. Even the same element from different vendors can handle different traffic loads. Thus, if the operator has multiple CDMA2000 radio vendors, then the operator will end up with multiple BTSs, BSCs, and MSCs that would need to be engineered accordingly. The configuration of AAA, HLR, VLR, and other back-office elements is also based on vendor implementation.

- *Backhaul (transmission) capacity:* The backhaul or transmission capacity depends on both bearer and signaling data that need to be transmitted between the network elements. Major connections include BTS-BSC, BSC-MSC,

BSC-PDSN/GGSN, MSC-back-office/PSTN, PDSN/GGSN-Internet. It also relies on the ATM transmission technology, which, for example, provides good multiplexing gain. The use of T1/E1 transmission lines with ATM as the transmission mode is common in the cellular world.

- *Coexistence with 2G (IS-95/GSM):* The migration from IS-95 to CDMA2000 was simple compared to that from GSM to W-CDMA. In most cases, this was because IS-95 base stations can be reused; CDMA2000 does not require a new spectrum; and CDMA2000 and IS-95 can be colocated in the base station. In contrast, WCDMA requires a newer base station and a new spectrum (2,100 MHz compared to 800/1,900 MHz). Thus, the transition to CDMA2000 was simpler and cheaper. However, in both circumstances, key aspects, such as intermodulation, guard bands, and handoffs, need to be carefully analyzed.

10.5.1.3 Network Optimization

After the construction of cell sites, calculations of cell capacity optimization are conducted. Network optimization consists of multiple steps that are described as follows:

1. *Baseline measurements:* The first step is to find out the current system performance that will serve as a baseline for the drive testing.
2. *Drive routes:* Proper planning of drive routes should be performed to verify coverage, capacity, and so forth. Drive routes should be planned for both cluster testing and system-wide testing.
3. *Manpower:* Test teams, tower crews, vendor support team, and so forth.
4. *Tools:* Diagnostic monitors, keys for log masks, spectrum analyzers, post-processing equipment and other RF test tools, and so forth.
5. *Documentation:* Detailed test plans, test scripts, and so forth.

The RF optimization process includes field tests conducted to tune and retune all aspects of CDMA air interface network performance. System-wide optimization is built on cluster testing and is an iterative process to fix coverage- and capacity-related issues. The key parameters that are evaluated include these:

- *Mobile receiving power:* The total power that the mobile is receiving from the BTS;
- *Mobile transmitting power:* The total power that the phone is transmitting;
- *Forward/reverse FER:* The percentage of forward/reverse FERs that is corrupted;
- E_c/I_o: The CDMA signal quality (defined as the CDMA chip energy divided by the present interference);
- *BER:* The bit error rate to understand the performance of data;
- *Service measurements:* Dropped calls, blocked calls, data rates, data throughputs, soft handoffs, and so forth.

Thus, areas of poor network performance are identified, and changes are implemented to fix the problems. These adjustments can include antenna changes, antenna tilting, antenna orientation, link budgets, and transmitted power. During testing, multiple vendors and different types of handsets are used to validate the performance of both handsets and the network. The testing is an iterative process that is repeated until the objectives of network planning are achieved; subsequently, the network is handed over to operations for live traffic testing and day-to-day network management.

10.5.1.4   Future Expansion Planning

The work of engineering teams does not end with network optimization. Because the number of subscribers can grow anytime, the engineering teams need to plan for the expansion of the network ahead of the curve. The subscriber forecast assists the engineering team in implementing the expansion. Two key options for expansion are cell splitting and carrier addition. Cell splitting adds multiple cell sites to aid an existing site with traffic handling. Carrier addition, on the other hand, adds extra carriers to the existing site to manage the growth of users. A third broad technique that is not common is the addition of new features (smart antennas, interference cancellation techniques) to make room for additional subscribers. The overall network planning process is shown in Figure 10.3.

10.6   Operations

10.6.1   Network Management

As service providers move into 3G and a mixed-supplier environment, they cannot afford separate management systems for the different network elements of different

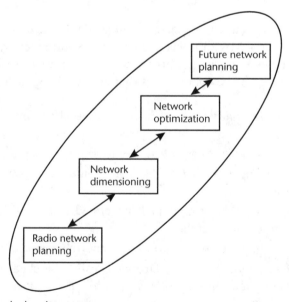

Figure 10.3   Network planning process.

suppliers. To address this issue, various standards groups within the ITU have defined interface and protocols for operations support systems under the umbrella of a telecommunications management network (TMN) [3].

A key to success for 3G1x-like networks is to minimize the cost of network operations. A good OAM&P platform will minimize the overall operations cost. With OAM&P systems service providers will be able to monitor network elements, add and remove equipment, test software and hardware, diagnose problems, and bill subscribers for services. The two key network management protocols are SNMP and CMIP:

- *SNMP:* The Simple Network Management Protocol (SNMP) was developed by the IETF for computer networks. It is not an efficient management protocol for wireless technologies such as 3G1x.
- *CMIP:* The Common Management Information Protocol (CMIP) was developed by ITU-T and OSI. It is designed to fulfill network management requirements in five functional areas as specified by TMN: fault management, accounting management, performance management, configuration management, and security management. CMIP is a better protocol than SNMP for cellular/PCS wireless networks.

For 3G, the management functions defined originally by TMN have to be expanded to cover beyond the five functional areas to support fraud management, software management, service management, roaming management, and customer profile management. The new TMN also needs to interface with different kinds of communications paths (e.g., dedicated lines, PSTN) to enable an efficient 3G network.

10.6.2 Device Management

Another important factor in 3G services is the device management element. Current cell phones are nothing more than dumb voice terminals; however, newer smart phones now provide access to the Internet, and they also load and run third-party applications. Service providers are working to develop tools to manage terminal and service mobility of such devices.

10.7 Summary

This chapter briefly highlighted the process from inception to deployment for wide-area wireless technologies. We described the initial steps of standardization of proof-of-concept trials by considering the 3G1x technology followed by the tedious and longer steps of deployment, which include radio network planning, optimization, and so forth. Last, we looked into the different aspects of network and device management.

Also, with the rapid advances in technology, operators have to constantly monitor and be prepared for the next big change in the network. Unlike the migration from 2G to 3G, which took 5 to 8 years, the migration from 3G to wireless data technologies is taking half of that time. For example, the current trend of migration from current voicecentric networks (3G1x) to datacentric technologies such as EV-DO only took 3 to 4 years. To migrate to EV-DO, the operators have to upgrade their existing base stations, OAM&P, and engineering systems and provide new devices. Thus, the migration from one wireless technology to another is keeping the different network organizations on the run.

References

[1]   Ojanpera, T., and R. Prasad, *Wideband CDMA for Third Generation Mobile Communications*, Norwood, MA: Artech House, 1998.

[2]   Lawrence, H., R. Levine, and R. Kikta, *3G Wireless Demystified*, New York: McGraw-Hill, 2002.

[3]   Garg, V. K., "Mangement of Personal Communications Services (PCS) Networks." In *Telecommunications Network Managment*, pp. 150–174, S. Aidarous and T. Plevyak (eds.), Piscataway, NJ: West Sussex, England: John Wiley and Sons, Ltd, IEEE Press, 1998.

CHAPTER 11

Wireless Access Security

11.1 Introduction

Wireless security is one of the hot topics in the communications industry. As more and more people are adopting wireless communications, it has become extremely important for the wireless service providers to protect their and their customers' assets. The element that is pushing the envelope is the growth of wireless data traffic on 3G wireless systems. With the advent of 3G technologies such as CDMA2000 and UMTS, operators have the ability to enhance their network security by using improved authentication and encryption algorithms. The key to a secure network is to provide end-to-end security solutions. Securing the wired portion of the network will not address the problem of air interface security and vice versa. Therefore, it is prudent to provide security for the overall network.

Because wireline communications have been around for a long time, they are fairly secure. For example, to eavesdrop on a wireline voice call, one would need access to the physical wires of the PSTN. The PSTN networks are relatively secure for voice calls. Similarly, in a cellular voice network the PSTN portion of the network is considered secure from the illegal sources. The MSCs, by design, usually have lawful interception capabilities, access to call data records, and secure measures against unlawful access.

On a similar note, wireline Internet has been secured with the incorporation of Web server applications (i.e., secure socket layer [SSL], transport layer security [TLS]), and operating systems have integrated virtual private networks (VPNs). Commonly used VPN solutions include the Point-to-Point Tunneling Protocol (PPTP), the Layer 2 Tunneling Protocol (L2TP), IP Security (IPsec), and SSL. These wireline security mechanisms are applicable to the wireless world and have been exhaustively tested and scrutinized by security experts worldwide, having proven to be robust and secure. Similar SSL and TLS technologies can be used in the wireless Internet systems, and VPNs can be employed to encrypt and protect the entire path.

Security for wireless systems has been mainly defined by the standards bodies for the radio access part of the network and its interactions with the devices and core wired networks. Therefore, our focus in this chapter will be on radio access network (RAN) security. The chapter first describes CDMA2000, UMTS, BWA, WLAN, and WPAN security measures. Then we briefly address the threats to access networks and possible mitigations. The last section provides an overview of an end-to-end (wireline and wireless) security.

11.2 CDMA2000 Access Security

The RAN defines the region between the handset and base station including the air interface, as shown in the Figure 11.1. Access security in cellular communication systems is composed of three main components: authentication, voice privacy/message integrity, and encryption. Authentication verifies that the device and network are who they say they are. Voice privacy protects conversations from eavesdropping, and message integrity ensures that the signaling information has not been modified in transit between the sender and the receiver. Encryption (confidentiality) guarantees the privacy of the signaling or transmitted data. This section briefly describes these three main components for CDMA2000 systems [1–3].

11.2.1 Authentication

Authentication is the first step in the overall security process. It is the process by which information is exchanged between a mobile station and base station for the purpose of confirming the identity of the mobile station. A successful outcome of the authentication process occurs only when it can be demonstrated that the mobile station and the base station possess identical sets of shared secret data.

Authentication needs to be completed before encryption can be initiated. IS-2000 Release B and earlier releases rely on cellular authentication and voice encryption (CAVE) to perform authentication, whereas Releases C and D can use authentication and key agreement (AKA) instead of CAVE to conduct verification. IS-856 Revision 0 relies on the Default Authentication Protocol, which does not provide any services except the transfer of packets between the Encryption Protocol and the Security Protocol. IS-856 Revision A can also use a secure hash algorithm (SHA-1) for authentication in addition to the Default Authentication Protocol.

11.2.1.1 IS-2000 (1xRTT and 1xEV-DV) Authentication

IS-2000 networks rely on a 64-bit A-Key and the electronic serial number (ESN) of the mobile for security. A random binary number called the random challenge shared secret data (RANDSSD), generated by the HLR, also plays a key role in the

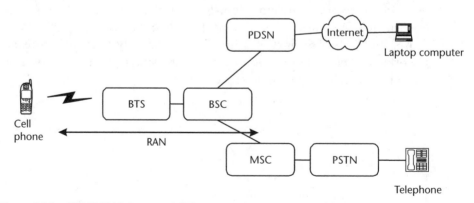

Figure 11.1 CDMA/EV-DO network [2].

authentication process. The A-Key is programmed into the mobile and is stored in the HLR.

IS-2000 uses the standardized CAVE algorithm to generate a 128-bit subkey called SSD. The A-Key, the ESN, and the network-supplied RANDSSD are the inputs to the CAVE that generate the SSD. The SSD is a 128-bit quantity that is stored in semipermanent memory in the mobile station and is readily available to the base station. The SSD has two parts: SSD_A (64 bit) for creating authentication and SSD_B (64 bit) for generating keys to encrypt voice and signaling messages. It can be shared with roaming service providers to allow local authentication. A fresh SSD can be generated when a mobile returns to the home network or roams to a different system.

In IS-2000 networks, the mobile uses the SSD_A and the broadcast random challenge (RAND) as inputs to the CAVE algorithm to generate an 18-bit authentication signature (AUTH_SIGNATURE) and then sends it to the base station. This signature is used by the base station to verify that the subscriber is legitimate. Both global challenge (in which all mobiles are challenged with the same random number) and unique challenge (in which a specific RAND is used for each requesting mobile) procedures are available to the operators for authentication. The global challenge method allows very rapid authentication. Also, both the mobile and the network track the call history count (a 6-bit counter). This provides a way to detect cloning, because the operator is alerted if there is a mismatch.

The 3G base stations can perform AKA authentication instead of the legacy verification with mobile stations. The purpose of AKA is to mutually authenticate the mobile station and the serving base station, and to establish a new integrity key (IK), encryption key (CK), and UIM[1] authentication key (UAK), if supported. On completion of the AKA, the CK and IK are stored in the mobile station while the UAK, if supported, is stored in the UIM. The details of the AKA procedure can be found in [4].

11.2.1.2   IS-856 (1xEV-DO) Authentication

Establishment of the initial connection is not authenticated but it does involve Diffie-Hellman key negotiation. The subsequent RAN domain messages can be authenticated and/or encrypted using the negotiated keys. The Default Authentication Protocol in IS-856 systems does not provide any services except for transfer of packets between the Encryption Protocol and the Security Protocol. This protocol does not register to receive any indications. On the other hand, the SHA-1 Authentication Protocol mainly verifies access channel MAC layer packets and does require a return indication.

11.2.2   Voice Privacy and Message Integrity

Voice privacy in IS-2000 systems is provided using a private long code mask (PLCM) that is used for PN. By design, all CDMA phones use a unique PN code for spreading the signal, which makes it difficult for the signal to be intercepted. The PLCM is utilized in both the mobile and the network to change the characteristics of

---

1.   The UIM is the user identity module capable of protecting the key.

the long code. This modified long code is used for voice scrambling, which adds an extra level of privacy over the CDMA air interface. The PLCM does not encrypt information; it simply replaces the well-known value used in the encoding of a CDMA signal with a private value known only to the mobile and the network. It is therefore extremely difficult to eavesdrop on conversations without knowing the PLCM.

The integrity of the message sent over the air is always protected. A message authentication code is calculated with a variant of HMAC-SHA-1 called keyed-hashing for message authentication (EHMAC).

11.2.3   Key Exchange and Encryption

Unlike integrity, confidentiality protection is applicable to both user and signaling data and voice. Two main components need to be considered when evaluating the security of an airlink (air interface) encryption method. The first is the algorithm itself and the second is the length of the key used by the algorithm to encrypt the signaling and data. In order for an unauthorized person to gain access to the encrypted information, the algorithm has to be broken or the secret key has to be determined.

An encryption algorithm takes as input unencrypted data and a key to generate encrypted data. Only authorized people who have the key can extract the data. The best way to prevent the algorithm from being broken is to openly publish the algorithm and rely on the secrecy of the key used for encryption to protect the data.

The second component to security is the length of the key used to encrypt and decrypt the data. For example, a key that is only 2 bits long can be easily broken because there are only four possible key values. However, a key that is 32 bits long has 2^{32} (more than 4.2 billion) possible values. On one hand, the key should be long, yet, on the other, it should be truly random. If there is any similarity or pattern in the bits in the key that can be exploited, the effectiveness of the key will be reduced. For instance, creating a 64-bit key by concatenating a 32-bit key with itself will not make a true 64-bit key. It will only generate an effective 32-bit key in terms of key strength.

11.2.3.1   IS-2000 Confidentiality

The mobile and the network use the 64-bit cellular message encryption algorithm (CMEA) key with the enhanced CMEA (ECMEA) algorithm to encrypt signaling messages sent over the air and to decrypt the information received. This 64-bit CMEA key is generated using SSD_B and the CAVE algorithm along with PLCM and a 32-bit data key. The 32-bit data key and an encryption algorithm called ORYX are used by the mobile and the network to encrypt and decrypt data traffic on the CDMA channels. Figure 11.2 illustrates the CDMA authentication and encryption mechanism.

The CDMA systems also support a temporary mobile station identifier (TMSI) for addressing the mobile station and it is assigned by the base station. This identity privacy feature makes it more difficult to correlate a mobile user's transmission to a mobile user. IS-2000 Release A and later have the option to use 128-bit Rijndael

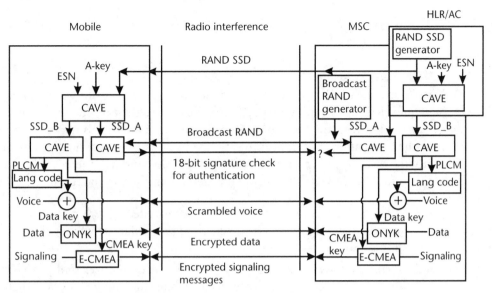

Figure 11.2   CDMA authentication and encryption [5].

algorithm, AES. AES is a block cipher that encrypts 128-bit blocks under the control of a 128-bit key.

11.2.3.2   IS-856 Confidentiality

The Default Encryption Protocol in IS-856 does not provide any encryption. The IS-856 has an option based on TIA-925 to use AES if required by the operator. IS-856-A can support AES.

11.2.4   Shortcomings in CDMA2000 Security

The key vulnerabilities of the CDMA2000 security architecture are as follows:

- Lack of support for public key technology, mainly because of lower performance results;
- No standardization of the transfer of the authorization vector, CK, and IK from the HLR/AC to PDSN or to VLR/MSC for circuit-switched services.

11.3   UMTS Access Security

UMTS includes all of the security measures that were part of GSM plus some additional ones to provide stronger security. It also enhances some of the existing GSM security mechanisms. Here are the key additional security elements of UMTS:

- Ciphering of both signaling information and user data;
- Security against use of false base stations with mutual authentication;

• Encryption that is extended from the air interface only to include Node-B and RNC connections.

The UMTS security aspects are specified in 3GPP specifications [6–8]. The UMTS specifications list five security feature groups as shown in Figure 11.3. The interfaces shown in Figure 11.3 provide certain security features. For example, the arrow between the mobile equipment (ME) and an access network emphasizes that the encryption and data integrity are provided bidirectionally. On the other hand, the direction of the arrow between the USIM and ME indicates that the some security features in USIM are kept confidential and should be received by ME.

1. *Network access security:* The set of security features that provide users with secure access to 3G services, which, in particular, protect against attacks on the (radio) access link.

2. *Network domain security:* The set of security features that enable nodes in the provider domain to securely exchange signaling data and that protect against attacks on the wireline network. This type of security is handled by the network operator or the equipment supplier.

3. *User domain security:* The set of security features that secure access to mobile stations.

4. *Application domain security:* The set of security features that secure applications in the user and in the provider domains.

5. *Visibility and configurability of security:* These features enable the user to determine whether a security feature is in operation or not and whether the use and provision of services should depend on the security feature.

The UMTS security architecture is based on three principles similar to CDMA2000: AKA, integrity protection, and confidentiality. A high-level view of these three security principles is shown in Figure 11.4 [9, 10].

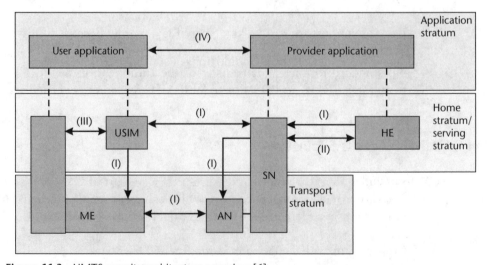

Figure 11.3 UMTS security architecture overview [6].

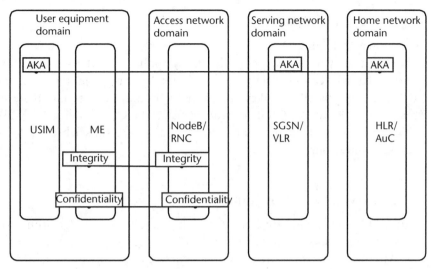

Figure 11.4   UMTS security principles [10].

11.3.1   Authentication and Key Agreement

The security architecture in UMTS is based on a mutual authentication procedure that is executed between the user (USIM) and the SGSN (packet switched) or MSC/VLR (circuit switched) at the network end. The procedure is called UMTS authentication and key agreement because, in addition to providing authentication service, it also includes generation of session keys for confidentiality and integrity protection at the user end. AKA is a two-staged process:

- Authentication of the user toward the network;
- Authentication of the network toward the user.

AKA is performed in so-called "one-pass authentication," reducing messages sent back and forth. After these procedures the user will be sure that he or she is connected to his or her served (trusted) network and the network is sure that the claimed identity of the user is true. Thus, the consequence of having mutual authentication is that the USIM is now an active entity. In GSM, the user could not authenticate the network; hence, the UE could not reject the network. Contrary to GSM, in UMTS the USIM will attempt to authenticate the network and it is now possible for the USIM to reject the network.

11.3.2   Integrity Protection

In UMTS, integrity protection is only provided for the signaling messages between the MS and the RNC. It is used to secure the content of the signaling messages between the user and the radio network even if the message is not confidential. If encryption is not available, then integrity protection is the only line of defense. Thus, the lack of user data integrity protection generates an issue in countries where encryption is not allowed or otherwise not available.

11.3.3 Confidentiality

As shown in Figure 11.4, encryption takes place between the MS and the RNC. Unlike integrity protection, confidentiality applies to both user data (voice and data) and signaling. Confidentiality is achieved by ciphering the user/signaling data between the subscriber and the access network and by referring to the subscriber by temporary identities (TMSI) instead of using the global identity (IMSI). If the network does not provide user data confidentiality, the subscriber is informed and can refuse connections.

11.3.4 UMTS Security Algorithms

The UMTS defines multiple cryptographic functions and algorithms to address the security concerns, as shown in Table 11.1. The cryptographic functions (f0–f5*) used in the AKA procedure are implemented exclusively in the USIM and authentication center (AuC). The UMTS operators are free to choose any algorithm but the recommended algorithm set for the AKA functions is MILENAGE. The cryptographic functions f0–f5* are, in principle, operator-specific functions, and there is no need for any interoperability of these functions between roaming partners. Functions f6 and f7 are specifically designed for MAP and network domain security. The KASUMI algorithm is the core algorithm used in functions f8 (confidentiality) and f9 (integrity). KASUMI is based on the block cipher (Misty) proposed by Mitsuru Matsui (Mitsubishi), first published in 1996. Misty translated from English to Japanese means KASUMI. KASUMI is a block cipher that produces a 64-bit output from a 64-bit input under the control of a 128-bit key. A detailed description can be found in [11]. The subscriber module USIM is issued by the HLR/AuC; the authentication-related security functions (f0–f5*) on the USIM have their counterpart in the AuC located in the HLR/AuC. Conversely, the encryption and integrity functions (f8 and f9) are located in the mobile device and the corresponding serving network (SN).

Table 11.1 UMTS Security Algorithms

Function	Description
f0	Random challenge generating function
f1	Network authentication function
f1*	Resynchronization message authentication function
f2	User challenge–response authentication function
f3	Cipher key derivation function
f4	Integrity key derivation function
f5	Anonymity key derivation function for normal operation
f5*	Anonymity key derivation function for resynchronization
f6	MAP encryption algorithm
f7	MAP integrity algorithm
f8	UMTS encryption algorithm
f9	UMTS integrity algorithm

11.3.5 Shortcomings in UMTS Security

The key deficiencies of the UMTS security architecture are very similar to CDMA2000 and are as follows:

- A major shortcoming is that the integrity of user data is not ensured.
- It lacks support for public key technology, mainly because of lower performance results.
- The transfer of authorization vector, CK, and IK from the HLR/AuC to SGSN or to VLR/MSC for circuit-switched services is not standardized.

11.4 BWA System Security

In this section, our focus will be mainly on IEEE 802.16 security. The IEEE 802.16 standard defines a separate *security sublayer* within the MAC layer to perform authentication, privacy, key exchange, and encryption. The security protocols described by the standard remain the same for the four different physical layers. As stated in the previous chapter, the MAC is connection oriented, which means that all transport and management connections have a connection ID. The IEEE 802.16-2004 standard secures only the transport and secondary management connections.[2] The IEEE 802.16-2004 security architecture has three components, described in the following subsections [12, 13].

11.4.1 Security Association

A security association (SA) is the set of security information that a BS and one or more of its client SSs share in order to support secure communications across the network. Three types of SAs are defined:

- *Primary SA*: SS establishes a primary security association during the SS initialization process.
- *Static SA*: Static SAs are provisioned within the BS.
- *Dynamic SA*: Dynamic SAs are established and eliminated in response to the initiation and termination of specific service flows.

The standard uses two types of SAs, data SA and authorization SA, but it has explicitly defined only data SA. The data SA protects transport connections between one or more SSs and a BS. The data SA consists of a 16-bit SA identifier or SAID, which is unique for each SA and data encryption standard (DES) in cipher block chaining (CBC) mode. It also provides two traffic encryption keys (TEKs) to encrypt data, TEK lifetime, 64-bit initialization vector for each TEK, and an indication of the type of data SA (primary, static, or dynamic).

The authorization SAs are used by base stations to configure data SAs on the SS. They consist of an X.509 certificate profile of the SS, 160-bit authorization key

---

2. Secondary management is used for delay-tolerant messages and for SS it carries IP management packets.

(AK), and AK lifetime. It also has a key encryption key (KEK), which is used by the BS to encrypt the TEKs and the downlink and uplink hash function-based message authentication code (HMAC) for data authenticity between the two elements.

11.4.2   Privacy Key Management

The Privacy Key Management (PKM) Protocol ensures the secure distribution of the keying data from the BS to SS. In addition, the BS uses the protocol to implement conditional access to the network services. PKM uses X.509 digital certificates, RSA public key encryption algorithms, and strong encryption algorithms to perform key exchanges between SSs and BSs. PKM applies the client–server model, where the SS acts as a client when requesting material from the BS, which acts as a server. PKM uses public key cryptography to derive a shared secret AK between the SS and the BS. The AK is used thereafter to avoid computationally intensive public key operations to derive the subsequent key operations.

A BS authenticates a SS during authentication exchange. Each SS carries its X.509 digital certificate issued by the SS's manufacturer, and RSA public/private key pairs provide internal algorithms to generate such key pairs dynamically. The digital certificate contains the SS's public key and the SS MAC address. When the BS receives an authorization request from the SS, the BS verifies the digital certificate and, if valid, generates an AK and encrypts it with the SS's public key and sends it back to the requesting SS. This authorization consists of a three-message exchange between a SS and a BS. The SS initiates the protocol by sending the first two messages, and BS responds with the third message.

11.4.3   Packet Data Encryption

The encryption services are defined as a set of capabilities within the MAC security sublayer. The encryption (DES-CBC) is only applied to the MAC PDU, but the generic MAC header (GMH) is not encrypted, as shown in Figure 11.5. The GMH carries 2 bits to indicate the TEK being used and does not carry the CBC mode initialization vector. The encryption module XORs the SA initialization vector with the contents of the PHY synchronization field from the most recent GMH to calculate the MAC PDU initialization vector. The MAC PDU initialization vector is quite predictable for two reasons. First, the SA initialization vector is constant and public for its TEK. Second, the PHY synchronization field is highly repetitive and therefore, predictable.

The DES-CBC does provide data encryption but no data authentication. To address data authentication, the IEEE 802.16-2004/.16e standards added the AES. The AES in CCM (counter with cipher-block-chaining MAC) mode is a new data link cipher that combines counter mode encryption for data confidentiality with the CBC-MAC for data authenticity. IEEE 802.11e also specifies AES in ECB mode to replace the Triple-DES (112-bit KEK) key wrapping in the PKM protocol.

11.4.4   Shortcomings in IEEE 802.16 Security

The IEEE 802.16e is expected to address the security deficiencies of the earlier version of the standard. The key weaknesses of IEEE 802.16-2004 are as follows:

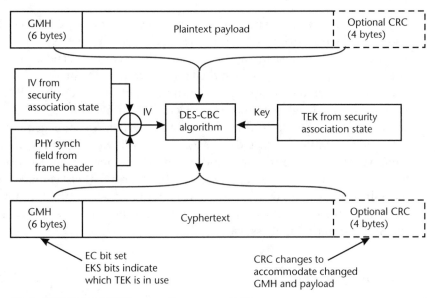

Figure 11.5   IEEE 802.16-2004 encryption process [12].

- An obvious flaw is the lack of mutual authentication or the absence of a BS certificate. The SS has to identify itself to the BS using its certificate, but the BS never has to prove its identity to the SS.
- DES in the CBC mode is not a powerful algorithm and has several security issues.
- There is no explicit definition for authorizing the SA.

11.5   WLAN Security

Since 1999, the IEEE 802.11 standard has provided sanctuary mechanisms to make WLAN security equal to that of its wired counterparts. The first attempt was wired equivalent privacy (WEP), which we all know has a number of flaws. After the failure of WEP and with the rise in popularity of WLANs, the 802.11 Task Group made another attempt to improve the security of WLAN networks. In June 2004, 802.11i was approved by the IEEE standards board; it enables better security architecture than WEP. In this section, we first take a look at WEP and 802.11i and then HIPERLAN2 [14–17].

11.5.1   IEEE 802.11-1999

The IEEE 802.11 standard introduced WEP to provide authentication, data integrity, and confidentiality. These processes are briefly described next.

11.5.1.1   Authentication

WEP has two authentication mechanisms: open system authentication and shared key authentication. In the former case, any STA that requests access to the network is granted and provided with a service set identity element (SSID). The SSID is used to identify the network to which this STA belongs. The open system authentication does not require WEP. In the latter case, the STA first sends an authentication request message to the AP, which then sends a random challenge. The challenge is a 128-octet generated by a WEP pseudorandom number generator (PRNG). The STA encrypts the challenge with its shared secret key, not specified in the standard. The secret key is distributed to the STAs by an external key management service. The STA sends back this encrypted message to the AP, which then verifies the accuracy of encrypted challenge response and finally sends back the authentication result to the station. If the authentication is successful, then a SSID is assigned to the station. The SSID is associated with a particular network so when the STA moves out of the coverage of the current network and enters another network, it needs to obtain a new SSID by initiating reassociation.

11.5.1.2 Confidentiality

After gaining access to the network, the stations can use WEP to encrypt messages. WEP used the Rivest Cipher 4 (RC4) stream cipher to encrypt messages using the shared secret key. The encryption process is shown in Figure 11.6, where encipherment begins with a secret key, a 24-bit initialization vector (IV), and the plaintext. WEP is a symmetric algorithm in which the same key is used for encipherment and decipherment. Initially, the secret key is concatenated with the IV, and the resulting seed is entered into a PRNG. The PRNG outputs a key sequence k equal in length to the plaintext–integrity check value (ICV) combination. The ICV, which is also known as CRC, is produced by applying an integrity algorithm on the plaintext. Then, finally key sequence k is mathematically combined with the plaintext-ICV block to produce a message that contains an IV and ciphertext.

The decryption process is illustrated in Figure 11.7. The secret key is concatenated with the received IV, which produces a seed. The seed is then passed to the PRNG to obtain the key stream. The key sequence is XORed with the ciphertext to obtain the plaintext and the ICV. The decipherment is verified by running the integrity check algorithm on the recovered plaintext and comparing the output ICV to the transmitted ICV. If the recovered and transmitted ICVs are not matched, then the received MAC PDU is assumed to have been tampered with and should not be passed to LLC.

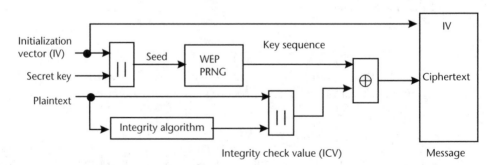

Figure 11.6 WEP encryption block diagram [16].

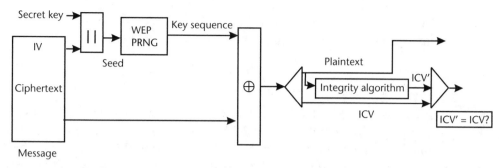

Figure 11.7 WEP decryption block diagram [16].

11.5.1.3 Vulnerabilities of WEP

WEP has failed to provide security in WLAN systems due to design flaws and because of poor implementations. Several successful attacks against WEP have been published and are widely available. The key vulnerabilities of WEP are as follows:

- CRC is an error detection code not a cryptographic function as it is used in WEP to offer integrity.
- The 802.11 standard does not specify how to generate and implement IVs. Because RC4 is a stream cipher, the same traffic key must never be used twice, but a 24-bit IV is not long enough to ensure this on a busy network.
- The 802.11 standard does not address how to distribute the keys or any key exchange protocol. The key has to be entered manually, which could be cumbersome in big networks. In some cases the system administrators have decided to use the same key and never change it, which further weakens the security of WLAN networks.

11.5.2 IEEE 802.11i-2002

The IEEE 802.11i standard was designed to address the shortcomings of WEP and provide an overall better security architecture. The 802.11i defines a robust security network (RSN), which, according to the standard, is a security network that only allows the creation of RSN associations. In other words, in an RSN the association between all stations including APs is built on a strong association/authentication called RSNA. The robust security network association (RSNA) is defined as an association used by a pair of STAs that depends on a four-way handshake for authentication or association. The standard also defined a framework that enables two classes of security algorithms for IEEE 802.11 WLANs:

- *Pre-RSNA:* A pre-RSNA security is made up of the IEEE 802.11 entity authentication and WEP algorithms.
- *RSNA:* RSNA security is made up of the Temporal Key Integrity Protocol (TKIP), Counter-Mode with CBC-MAC Protocol (CCMP), and 802.1X algorithms and enhance key management procedures.

In this section, we illustrate the authentication, key management, and confidentiality mechanisms that are defined in the 802.11i standard.

11.5.2.1 Improved Authentication

A RSNA utilizes 802.1X for its authentication and key management services. It incorporates two components: 802.1X port and authentication server (AS). There is one-to-one mapping between the 802.1X port and the association. The 802.1X port only allows traffic to pass through when the authentication is successfully completed. The 802.1X port consists of one controlled port and one uncontrolled port. The controlled port does not allow any traffic to pass through until it is cleared by an 802.1X authentication procedure that is conducted through the uncontrolled port. The AS could be a stand-alone server or it could be integrated with an access point. The protocol between the AS and AP is not specified by 802.11i, but it could use TLS or IPSec to provide secure communications between the two elements.

802.1X uses the Extensible Authentication Protocol (EAP) to authenticate stations and the server. It defines a generalized framework for multiple authentications, so a particular application could use the EAP framework and authenticate its users using any authentication method.

11.5.2.2 Key Management

The 802.1X defines two new key distribution techniques to provide authentication and confidentiality: four-way handshake and group key handshake. The four-way handshake is a pairwise key management protocol that requires mutual possession of a pairwise master key (PMK) by two parties and distributes a group temporal key (GTK). The group key handshake is a group key management protocol used only to issue a new group temporal key (GTK) to peers with whom the local STA has already formed security associations.

11.5.2.3 Enhanced Confidentiality

RSNA supports the TKIP and CCMP protocols for data confidentiality and data integrity. CCMP is mandatory in all IEEE 802.11 devices claiming RSNA compliance, whereas TKIP is optional for an RSNA. A design goal for TKIP was that the algorithm should be implemented within the capabilities of most devices supporting only WEP, so that many such devices would be field-upgradable by the supplier to support TKIP.

Like WEP, TKIP uses the RC4 algorithm, but it is implemented differently to address the vulnerabilities of WEP. The standard extended the IV to a 48-bit number to avoid replay attacks. It also added a new hash function MIC (Michael), which is keyed to ensure data integrity. The CCMP is more secure than TKIP because it uses AES and also provides authentication, confidentiality, integrity, and replay protection. The details of the TKIP and CCMP protocols can be found in [17].

11.5.3   HIPERLAN Type 2 Security

HIPERLAN2 security is similar to that provided by IEEE 802.11-based WLANs. The security architecture has two main functions, encryption and authentication, whereas key management is a supporting function for both encryption and authentication [18, 19].

- *Authentication:* HIPERLAN2 supports mutual authentication, which means that STA has to verify AP and AP has to authenticate STA. The STA is assigned an authentication key identifier that needs to be presented to the connected AP at connection. The identifier is protected by encryption during transfer to the AP. There are six possible types of identifiers for the authentication key: 48-bit IEEE address, 64-bit extended IEEE address, network access identifier, distinguished name, compressed type, and generic type. The details can be found in [19]. In addition, HIPERLAN2 supports five alternatives for the authentication protocol: preshared key, RSA512, RSA768, RSA1024, and no authentication.

- *Key management:* Key management consists of key generation and refresh as well as handling of keys to external entities such as key databases. The keys used for authentication are long-term keys, available on both the STA and AP sides even before authentication (if desired) can take place. Short-term keys are used for unicast and broadcast/multicast encryption and can be refreshed. The interval for key refresh is regulated by the local security policy. The unicast secret session key (SSK) is a secret key known only to one MT and its connected AP, and it is valid for a limited period. Common keys are used for encryption of multicast and broadcast user data and are used if encryption is chosen in the join procedure.

- *Encryption:* Encryption provides confidentiality to the data. HIPERLAN2 uses two encryption algorithms, DES and Triple-DES, in order to support different security levels. These algorithms use an IV for encryption and decryption. The implementation of DES is mandatory, whereas the use of Triple-DES is optional. A mapping between the encryption key for DES, as well as the ones for Triple-DES and a MAC ID, is defined in [19].

11.5.4   WLAN Security Analysis

The IEEE 802.11i standard was designed to address many weaknesses of the previous versions of the standard. To enhance encryption, it supports TKIP and CCMP (AES) in addition WEP. WEP has number of shortcomings, but those can be avoided by implementing 802-11i mechanisms. The original standard does not define a key distribution protocol. The 802.11i uses the four-way handshake and group key handshake to provide authentication and confidentiality.

11.6   IEEE 802.15.1 (Bluetooth) Security

Bluetooth, like other WPAN technologies, provides peer-to-peer communications over a short distance. To provide usage protection and information confidentiality, the system has to provide security measures both at the application layer and the link layer. In this section, we mainly focus on the link security where four entities have been defined to maintain security [20, 21]:

- *Bluetooth device address:* Each Bluetooth device has a publicly known unique identification number, a 48-bit Bluetooth address (BD_ADDR).
- *Private authentication key:* This is a 128-bit key used in the authentication processes.
- *Private encryption key:* This has a variable length of 8 to 128 bits, depending on the level of the security needed for the application.
- *RAND:* This is a 128-bit random number that is different for each new transaction. The Bluetooth devices use this RAND for various purposes in the encryption and authentication processes.

In the next several paragraphs, our focus will be on the three key aspects of Bluetooth security: key management, encryption, and authentication.

11.6.1 Key Management

Key management is a crucial item to consider due to use and types of keys defined in Bluetooth. The authentication key is often referred as the *link key* since it is specific to a Bluetooth link. In addition to link keys, there is also an encryption key.

11.6.1.1 Link Key

The link key used in the authentication process is a 128-bit random number that is shared between two or more devices. The link key is also used as one of the parameters from which the encryption key is derived. It could be either semipermanent or temporary. The semipermanent key is stored in the nonvolatile memory and can be used after the session is terminated for the other sessions. The temporary key, on the other hand, is limited by the lifetime of the current session; it cannot be reused in a later session. Depending on the type of application, link keys can be initialization keys, unit keys, combination keys, or temporary (master) keys as described here:

- *Initialization key, K_{init}:* The initialization key is used during the authentication process when no other keys are available. It is required when two devices, with no prior communications, need to talk to each other.
- *Unit key, K_A:* The unit key is generated in a single device when it is installed. After its generation, it is stored in the nonvolatile memory but is rarely used.
- *Combination key, K_{AB}:* The combination key is derived from two devices, and it is generated for each new pair of Bluetooth devices.
- *Temporary (master) key, K_{master}:* The master key replaces the current link key and can be used when the master unit wants to transmit information to more than one recipient. It is valid only during the current session.

11.6.1.2   Encryption Key, K_c

This key is derived from the current link key. It is shorter in length compared to the link/authentication key to facilitate different encryption options without weakening the authentication key.

11.6.2   Encryption

Encryption protects the user information by encrypting the packet payload; it does not encrypt the packet header. The encryption of payloads is carried out with a steam cipher called E_0, which is resynchronized for every payload. The E_0 stream cipher consists of the three parts—the payload key generator, the key stream generator, and an encryption/decryption part—as shown in Figure 11.8.

The payload key generator combines the inputs (K_c, address, master clock, and RAND) bits and shift them into the four linear feedback shift registers (LFSRs) used in the key stream generator. The key stream bits are generated by a method derived from the summation stream cipher generator attributable to Massey and Rueppel. The generated key stream cipher is modulo-2 added to the data to be encrypted. The cipher is symmetric; decryption is performed using the same key.

Depending on whether a device uses a unit/combination key or a master key, several encryption modes are available. If a master key is used, there are three possible modes, unlike for the unit/combination key, which is typically used for more than one session and is potentially less secure. In encryption mode 1, nothing is encrypted. In encryption mode 2, broadcast traffic is not encrypted, but the individually addressed traffic is encrypted with the master key. In encryption mode 3, all traffic is encrypted with the master key.

11.6.3   Authentication

The authentication mechanism is based on a challenge–response scheme, in which a two-move protocol is used to check if the other party knows the secret key, as shown in Figure 11.9.

The protocol uses symmetric keys, so a successful authentication is based on the fact that both participants share the same key. In the challenge–response scheme, the verifier challenges the claimant to authenticate a random input (the challenge), denoted by AU_RANDA, with an authentication code, denoted by E_1. The claimant then returns the resulting signed response (SRES) to the verifier for verification. If the authentication fails, a certain period of time must pass before a new attempt can be made. The period of time doubles for each subsequent failed attempt from the

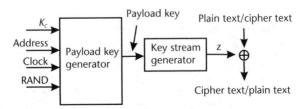

Figure 11.8   Stream cipher E_0 block diagram [21].

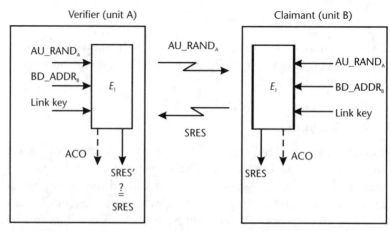

Figure 11.9 Challenge response for Bluetooth [21].

same address, until the maximum waiting time is reached. The waiting time decreases exponentially to a minimum when no failed authentication attempts are made during a time period.

11.6.4 Bluetooth Security Analysis

Bluetooth security has it own limitations:

- Bluetooth only offers device authentication and not user endorsement.
- The unit key is vulnerable to attacks from insiders.
- It only provides link level security; it does not provide security at the application level.

11.7 Access Network Security Threats and Possible Mitigations

The two major network resources that are part of the access network are BTS/Node-B and BSC/RNC and AP beside the medium of free space. The three primary areas around RAN security are paging, access, and traffic channels. Some of the possible RAN security threats and possible mitigation to those threats are shown in Table 11.2.

11.8 Other Threats and Mitigations

11.8.1 Phone Fraud and Theft of Service

Although fraud and theft of service are not directly related to RAN, it is worthwhile to briefly look into these two aspects.

Phone fraud has been part of the telephone system since the very beginning, whether in the form of the consumer attempting to defraud the telephone company, the telephone company attempting to defraud the consumer, or a third party

Table 11.2 RAN Security Threats and Possible Mitigation Techniques

Category	Threats	Possible Mitigation
Air interface denial of service (DoS) attacks	Paging channel flooding Generation of excessive access attempts Generation of excessive traffic (theft of bandwidth)	Secure software and hardware platforms Firewalls
BSC/RNC/AP DoS threats	Mass sources of connection setup and teardown: possible scaling issue with numerous packets required between BSC/RNC and PDSN for each setup/teardown event Malicious signaling injection	Firewalls
Confidentiality threats (snooping of messages)	Paging channel customer data visibility (unencrypted data) Attacks on encryption algorithms (if available) Thefts of services and credentials Eavesdropping	Authentication and encryption of paging, control, and access channel messages
Data integrity threats	Malicious injection of data	Firewalls

attempting to defraud either of them. The phone companies are committing fraud or cramming if they charge their subscribers for services that were neither ordered nor desired by the client or for fees for calls or services that were not properly disclosed to the consumer. The telemarketing fraud committed by third parties involves solicitations for the sale of goods or investments that are never delivered or worthless and requests for donations to bogus, unregistered charities. The users will be charged with subscription fraud if they sign up with a bogus name.

Theft of services is the legal term for a crime that is committed when a person obtains valuable services as opposed to goods by deception, force, threat, or other unlawful means, that is, without lawfully compensating the provider of said services. Laptop theft is a serious threat to users of mobile computers where victims can lose hardware, software, and personal and office data to thieves. Data theft is another growing issue that is primarily perpetrated by office workers with access to technologies, such as handheld devices, that can store big chunks of data [22].

11.8.2 Intrusion Detection

Intrusion detection is the process of identifying and responding to malicious activity targeted at computing and networking resources. Unlike firewalls, which are the first line of defense, intrusion detection systems (IDSs) come into the picture only after an intrusion has occurred and a node or network has been compromised. That is why IDSs are aptly called the second line of defense. The malicious acts include but are not limited to network attacks against vulnerable services, data-driven attacks on applications, host-based attacks such as privilege escalation, unauthorized logins and access to sensitive files, and malware (viruses, Trojan horses, and worms) [22, 23].

Intrusion detection can be classified into three broad categories:

- *Anomaly detection:* In an anomaly detection system, a baseline profile of normal system activity is created. Any system activity that deviates from the baseline is treated as a possible intrusion.
- *Misuse detection:* Such a system tries to detect evidence of intrusive activity irrespective of any knowledge regarding the background traffic (i.e., the normal behavior of the system).
- *Specification-based detection:* Specification-based detection defines a set of constraints that describe the correct operation of a program or protocol, and monitors the execution of the program with respect to the defined constraints. This technique may provide the capability to detect previously unknown attacks, while exhibiting a low false-positive rate.

The type of intrusion response for wireless networks depends on the type of intrusion, the network protocols and applications in use, and the confidence (or certainty) in the evidence. Two likely responses include reinitializing communication channels between nodes and notifying the user who in turn takes appropriate action.

11.9   End-to-End Network (Wireline and Wireless) Security Analysis

Figure 11.10 illustrates high-level topologies for 3G cellular and wireless WAN, PAN, and BWA systems. In the earlier sections of this chapter, our focus was on securing access to part of these various wireless systems. As we all know, however,

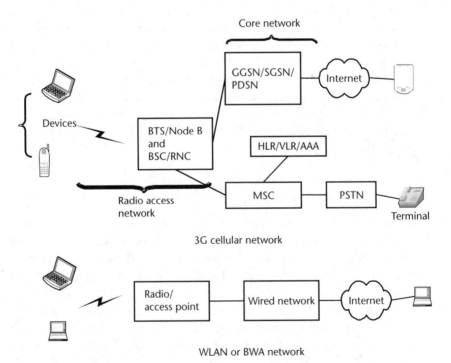

Figure 11.10   End-to-end security for wired and wireless networks.

network operators have to provide end-to-end security, so securing one important piece will not be prudent. At a high level, the figure has three main components: the device, the radio access (including air interface) network, and the core wired network. The standards mainly provide specifications to secure the RAN and its interactions with the devices and core network. The operators have to work with the rest of the software and hardware suppliers to enable security mechanisms for devices and the core network.

- *Device:* First and foremost, the cellular phones/stations need protection from viruses and worms, which are on the rise. The increase in the number of smart phone devices, based on the Palm, Linux, Microsoft, and Symbian operating systems, is pushing the envelope. Smart phones have many things in common with desktops and laptops and, most interestingly, they both can load and run third-party destructive applications. Although the smart phones market is comparatively very small, it is on the rise, which makes them more vulnerable to attacks and adds to the anxiety of operators.[3] Thus, operators and handset manufacturers are looking to operating system, antivirus, and semiconductor suppliers for possible solutions.

- *Core wired network:* The core network, which includes MSCs, PDSNs, SGSNs/GGSNs, and other key databases, has been made relatively secure with the adoption of firewalls and other network-based solutions, but improvements are still needed. These solutions provide the ability to monitor network traffic, eliminate spam, and screen text messages for viruses before they can enter the domain of the operator.

- *SS7 and SIGTRAN:* The MAP and IS-41 employ the circuit switching of the SS7 protocols for data transport in GSM and CDMA, respectively. MAP and IS-41 messages are exchanged between MSCs and HLRs and SGSNs and HLRs via the SS7 domain. The Signaling Transport (SIGTRAN) Working Group provided security mechanisms to protect the integrity of signaling information. These mechanisms are provided to support SS7 signaling over IP networks (such as the Internet). SIGTRAN recommends the use of IPsec, which ensures authentication, confidentiality, and data (signaling) protection. In IPSec a key concept is security association, which is a one-way relationship between the sender and the receiver and offers security services to IP traffic [24].

- *VPN and IPsec:* The use of VPN tunneling and IPsec as a transport protocol will make the network more secure and robust to attacks. The VPN gateway provides authentication, data integrity, and privacy. In the case of WLANs in particular, all wireless traffic must go through the VPN server before access to the larger wired network is granted. The use of VPN solutions will enable the mobile workforce to be more effective.

- *Interworking security:* As stated earlier, UMTS and WLAN have specified AKA and EAP to provide network access security. Thus, when a 3G user wants to access a WLAN network, the authentication traffic always need to return to the 3G core network through WLAN. This results in an authentica-

---

3.   The term *operators* includes mobile, fixed (BWA), WLAN, and WPAN providers.

tion delay, which is not acceptable for real-time traffic. Thus, a localized authentication mechanism is needed to minimize the authentication delay. In addition, EPA-AKA reveals user's identity to WLANs, which is not acceptable to the subscriber. Another issue is how to carry out mobility and AAA procedures, while considering both IP and link layer security across administrative domains. It is also important to minimize security procedures during handovers [25]. The details of the 3G and WLAN interworking security can be found in [26].

11.10   Summary and Future Work

Currently, many deployed networks and newer handsets/stations are capable of performing encryption and authentication, but those have not been turned on for one reason or another. Even if authentication and data integrity are allowed, the keys used are still derived primarily from older procedures. It will be a few years before the vast majority of the equipment is capable and derives full-strength keys (e.g., 128-bit keys, AES). It is particularly unfortunate that many networks are running without any authentication and encryption at all. Finally, the authentication/encryption has to be enabled at all levels—higher level, core network, radio, and device—to truly facilitate end-to-end security.

References

[1]   Rose, G., and G. Koien, "Access Security in CDMA2000, Including A Comparison with UMTS Access Security," *IEEE Wireless Communications*, Vol. 11, No. 1, February 2004, pp. 19–25.

[2]   Qualcomm, "Comparison of Airlink Encryptions," 2003.

[3]   Qualcomm, "1xEV-DO Security," 2003.

[4]   "Upper Layer (Layer 3) Signaling Standard for CDMA2000 Spread Spectrum Systems," Standards & Technology Department, TIA-2000.5-D, TIA, March 2004.

[5]   Winget, C., and M. Naidu, "CDMA 1xRTT Security Overview," *Qualcomm*, August 2002.

[6]   "3G Security; Security Architecture," Technical Specification (Release 6), Technical Specification Group Services and Systems Aspects, 3GPP TS 33.102 (V6.4.0), 3GPP, September 2005.

[7]   "3G Security; Security Principles and Objectives," Technical Specification (Release 4), Technical Specification Group Services and Systems Aspects, 3GPP TS 33.120 (V4.0.0), 3GPP, Mar. 2001.

[8]   "3G Security; Security Threats and Requirements," Technical Specification (Release 4), Technical Specification Group Services and Systems Aspects, 3GPP TS 21.133 (V4.1.0), 3GPP, Dec. 2001.

[9]   Koien, G., "An Introduction to Access Security in UMTS," *IEEE Wireless Communications*, Vol. 11, No. 1, February 2004, pp. 8–18.

[10]   "UMTS Security Features," Technical Brief, Tektronix, 2004.

[11]   "3G Security; Specification of the 3GPP Confidentiality and Integrity Algorithms; Document 2 KASUMI specification," Technical Specification (Release 6), Technical Specification Group Services and Systems Aspects, 3GPP TS 35.202 (V6.1.0), 3GPP, September 2005.

[12]   Johnston, D., and J. Walker, "Overview of IEEE 802.16 Security," *IEEE Security and Privacy*, Vol. 2, No. 3, May–June 2004, pp. 40–48.

[13] "IEEE Standard for Local and METROPOLITAN AREA NETWORKS, Part 16: Air Inter-face for Fixed Broadband Wireless Access Systems," IEEE Std 802.16–2004, IEEE, June 2004.

[14] Chen, J., M. Jiang, and Y. Liu, "Wireless LAN Security and IEEE 802.11i," *IEEE Wireless Communications*, Vol. 12, No. 1, February 2005, pp. 27–36.

[15] Smith, C., and J. Meyer, *3G Wireless with WiMax and Wi-Fi: 802.16 and 802.11,* New York: McGraw-Hill, 2004.

[16] "IEEE Standard for Information Technology—Telecommunications and Information Exchange Between Systems—Local and Metropolitan Area Networks—Specific Require-ments—Part 11: Wireless LAN Medium Access Control (MAC) and Physical Layer (PHY) Specifications," IEEE Std 802.11, IEEE, 1999.

[17] "IEEE Standard for Information Technology—Telecommunications and Information Exchange Between Systems—Local and Metropolitan Area Networks—Specific Require-ments—Part 11: Wireless LAN Medium Access Control (MAC) and Physical Layer (PHY) Specifications; Amendment 6: Medium Access Control (MAC) Security Enhancements," IEEE Std 802.11i, IEEE, 2004.

[18] "Broadband Radio Access Networks (BRAN); HIPERLAN Type 2; Data Link Control (DLC) Layer; Part 1: Basic Data Transport Functions," Technical Specification, ETSI TS 101 761-1 (V1.3.1), ETSI, December 2001.

[19] "Broadband Radio Access Networks (BRAN); HIPERLAN Type 2; Data Link Control (DLC) Layer; Part 2: Radio Link Control (RLC) Sublayer," Technical Specification, ETSI TS 101 761-2 (V1.1.1), ETSI, April 2000.

[20] Hager, C. T., and S. F. Midkiff, "An Analysis of Bluetooth Security Vulnerabilities," *Proc. IEEE Wireless Communications and Networking Conf.*, New Orleans, LA, March 16–20, 2003, pp. 1,825–1,831.

[21] "IEEE Standard for Information Technology—Telecommunications and Information Exchange Between Systems—Local and Metropolitan Area Networks—Specific Require-ments—Part 15.1: Wireless Medium Access Control (MAC) and Physical Layer (PHY) Specifications for Wireless Personal Area Networks (WPANs)," IEEE Std 802.15.1, IEEE, 2002.

[22] Wikipedia, http://en.wikipedia.org/wiki/Main_Page.

[23] Mishra, A., K. Nadkarni, and A. Patcha, "Intrusion Detection in Wireless Ad Hoc Net-works," *IEEE Wireless Communications*, Vol. 11, No. 1, February 2004, pp. 48–60.

[24] Young, M., et al., "Transport of Mobile Application Part Signaling over Internet Protocol," *IEEE Communications Magazine*, Vol. 40, No. 5, May 2002, pp. 124–128.

[25] Yang, C., K. Chu, and Y. Yang, "3G and WLAN Interworking Security: Current Status and Key Issues," *Int. J. of Network Security*, Vol. 2, No. 1, January 2006, pp. 1–13.

[26] "3G Security; Wireless Local Area Network (WLAN) Interworking Security," Technical Specification (Release 7), Technical Specification Group Service and System Aspects, 3GPP TS 33.234 (V7.1.0), 3GPP, June 2006.

CHAPTER 12

Wireless Communications Evolution: Beyond 3G

12.1 Introduction

As I was writing this chapter, the world of wireless communications continued to move beyond 3G. Unfortunately, the activities within different standards bodies are still shifting away from the goal of one single mobile wireless communications standard across the globe. This goal was set up by the IMT-2000 committee for the convergence of various 2G communications, but it never got fulfilled within 3G communications.

At this moment there are three different standard bodies working on three different but similar (in terms of technicalities) standards. These standards are EV-DO Revision C, IEEE 802.20, and 3G-LTE. They are currently under development by the 3GPP2, IEEE, and 3GPP committees, respectively. We can also include IEEE 802.16e in the mix, which was completed in December 2005. But the WiMAX Forum has to go one step further, because the IEEE standard only defines the lower physical and MAC layers; it has yet to define the compelling network architecture. We are considering these four standards[1] to be 3.5G technologies and hoping that in the next few years the standards bodies will bury their differences, converge, and support one unified mobile wireless communications standard, that is, 4G, in the long run.

The focus of this chapter will be on the 3G/3.5G standards (EV-DO Revision C, IEEE 802.20, and 3G-LTE), key enhancements/technologies for 3.5G and 4G technologies, and convergence aspects of cellular and IEEE-based technologies on the basis of a single 4G standard.

12.2 Current Activities

Before we dive into the different aspects of 3.5G and 4G technologies, let's briefly review the activities that are currently happening around the 3G globe:

- *3GPP2*: The majority of the CDMA2000 1X and EV-DO Revision 0 deployments are in North America and Asia Pacific; markets in the rest of the world are starting to catch up. There are about 225 million and 24 million CDMA2000 1X and 1xEV-DO subscribers, respectively, in the world [1].

---

1. The 802.16e technology can be considered a 3.5G technology when it truly enables mobility, which is not currently supported.

Some of the key operators are now planning to launch EV-DO Revision A in the next 1–2 years.

- *3GPP:* GSM has, by far, the largest subscriber base and it is still growing. The majority of the GSM/GPRS and WCDMA networks are in Asia Pacific and Europe. There are more than 1.5 billion GSM/GPRS and 40 million WCDMA subscribers in the world [2]. Even though the WCDMA market is growing slowly, it has been able to capture around 100 countries. HSDPA is around the corner and HSUPA is expected to follow along.

- *CCSA:* China's Ministry of Information Industry is expected to issue 3G licenses in 2007. It is likely that the government will only issue one 3G/TD-SCDMA license initially, while other 3G licenses will be awarded at a later date. The deployment of TD-SCDMA outside China is highly unlikely at least for now [3].

- *IEEE:* WiMAX (802.16e) PHY and MAC layer specifications were completed in December 2005 and the WiMAX Forum is working to complete the upper layer specifications. The 802.16d-certified devices are still very expensive compared to CDMA2000 and WCDMA handsets. Very few 802.16d FWA networks have been deployed to this point, but the market is expected to grow as a replacement for DSL. The IEEE 802.20 (rival to WiMAX and disbanded for the time being) committee is also working at a fast pace to complete an alternative mobile broadband standard.

12.3   3GPP2: Beyond 3G

The two key proposals submitted for the evolution of CDMA2000 family are EV-DO Revision B and EV-DO Revision C.

12.3.1   EV-DO Revision B

EV-DO Revision B was standardized in early 2006. Its most significant enhancement is the concatenation of the 1.25-MHz carriers (scalable bandwidth). It allows aggregation of up to fifteen 1.25-MHz carriers within 20 MHz of bandwidth. Revision B introduces a 64-QAM modulation scheme and will deliver peak rates of 73.5 Mbps in the forward link and 27 Mbps in the reverse link within 20 MHz of bandwidth. A single 1.25-MHz carrier in the downlink can deliver a peak rate of up to 4.9 Mbps. In addition, the diversity and interference cancellation schemes are introduced to support mobile broadband data. All of these enhancements will improve the performance of delay-sensitive applications such as VoIP, video telephony, multimedia, and multiplayer online gaming [1].

12.3.2   EVDO-C

Qualcomm, Nokia, and a collaboration of Lucent, Nortel, and Samsung (LNS) teams have submitted three key but similar proposals for the evolution of the CDMA2000 air interface. We highlight the key aspects of Qualcomm and LNS proposals in this section, one of which is expected to be approved in early 2007.

The LNS proposal supports a 1.25-MHz up to 20-MHz carrier bandwidth using the TDD or FDD mode. The performance objectives (peak date rate and aggregate sector throughput) are set at 150 to 200 Mbps and 75 Mbps within a 20-MHz channel, respectively. Also proposed are two modes of operation for forward and reverse links: the loosely backward compatibility (LBC) mode and the strictly backward compatibility (SBC) mode. In LBC, legacy terminals and new terminals have to operate on separate carriers. The downlink is based on OFDMA, whereas the uplink uses a hybrid of OFDMA and CDMA carriers. In SBC mode legacy, new and legacy terminals exist on the same carrier. Both the downlink and uplink are based on a hybrid OFDMA/CDMA scheme, whereas the uplink also supports CDMA on legacy carriers. The proposal also includes support for technologies such as MIMO, SDMA, and interference cancellation [4].

Qualcomm's proposal supports a fully (strict) backward compatible evolution of the EV-DO Revision B standard. The forward link slot occupies a 1.6384-MHz channel, which consists of OFDM symbols and guard tones but has the same number of chips as EVDO-A. The downlink also supports a unified framework for MIMO and SDMA. Closed-loop MIMO enables beam-forming gain to users with multiple transmitting antennas where the beams are selected from within the predefined code book. The SDMA, on the other hand, provides subscribers with access to the communication channel based on their spatial locations. Qualcomm also proposes a superposition coding technique to compensate for the inefficiencies of the TDM-based systems. In the uplink, the proposal indicates support for advanced receivers capable of linear and nonlinear interference cancellation in the CDMA waveform. The proposal also brought back the support for the 3G1x features of active-hold state and quick paging, which can be used to reduce connection times and extend battery life [5].

12.4   3GPP: Beyond 3G

For more than a year, the 3GPP has been working to define requirements for systems beyond 3G. In 3GPP terminology, the future "beyond 3G" is commonly referred to LTE (long-term evolution) for the radio access network and SAE (system architecture evolution) for the network architecture. The goal was to finish the study by June 2006 with all relevant core specifications by mid 2007.

12.4.1   3G-LTE

The 3G-LTE has more developed requirements than SAE. The LTE of the RAN (Release 8) has several performance objectives, including these:

- A downlink peak data rate of 100 Mbps within a 20-MHz spectrum;[2]
- An increased downlink average user throughput and spectrum efficiency that is three to four times greater than that of Release 6;

2.   To achieve the full benefits of 3G-LTE, operators will need a 20-MHz bandwidth, which will require new spectrum allocations such as the GSM extension band (880–890 and 925–935 MHz), for which they will experience strong competition.

- An uplink peak data rate of 50 Mbps within a 20-MHz spectrum;
- An uplink average user throughput and spectrum efficiency that is two to three times greater than that of Release 6;
- Support for different spectrum allocations in multiples of 1.25 MHz up to 20 MHz;
- A minimum of 200 users per cell in the active state in a 5-MHz bandwidth;
- Latency of as little as 20 ms in some instances;
- Full mobility at speeds from 0 to 350 km/h and possibly 500 km/h;
- Support for an overlay on existing UMTS networks;
- Support for handovers between LTE and 3G.

To address these physical layer objectives, the working groups have identified six proposals for further study:

1. Nokia, Ericsson, and others submitted a proposal for OFDMA in the downlink and SC-FDMA in the uplink (FDD mode).
2. Nortel submitted a proposal for bidirectional OFDMA (FDD mode).
3. Qualcomm submitted a proposal for MC-WCDMA (FDD mode).
4. The Chinese Academy of Telecommunications Technology and Datang submitted a proposal that uses TD-SCDMA.
5. IP Wireless submitted a TDD/FDD proposal that uses TD-OFDM.
6. Samsung submitted a proposal using OFDMA in the uplink and SC-FDMA in the downlink based on the TDD mode (WiBro).

12.4.2   System Architecture Evolution

The details around SAE are still being hashed out and decisions were expected by the end of 2006. The goal is to develop an architecture that can support multiple radio access technologies including non-3GPP and WLAN. The primary focus is on the packet-switched domain with the assumption that VoIP will be supported.

The SAE is also expected to converge the five primary nodes (Node-B, RNC, MSC, GGSN, and SGSN) into two nodes: an enhanced Node-B, which includes some RNC functionality, and a single core network node [6].

12.5   IEEE 802.20

Through the IEEE 802.20 or MBWA working group has been disbanded for the time being, it was revitalized in January 2006 via a proposal from Qualcomm. The proposal was in response to the IEEE 802.16e (WiMAX) standard. It defines the PHY and MAC layer architectures for the air interface while the rest of the protocol stack is based on the EV-DO technology. It consists of access terminals, access network, and HA-like EV-DO. Similar to EV-DO Revision C, the peak data rate is more than 130 Mbps in a 20-MHz channel with a scalable bandwidth of 5 to 20 MHz. The proposal is expected to be approved in 2007 [7].

The proposal supports two modes of operation for the lower MAC sublayer and physical layer: single input and single output (SISO) and MIMO. The MIMO mode is divided into two submodes: multiple codeword (MCW) and single codeword (SCW). The downlink is based on OFDMA with MIMO support and also includes precoding and SDMA features. In the uplink, orthogonal transmissions are based on OFDMA and nonorthogonal transmissions are supported with multiple antennas.

12.6 Key Enhancement Technologies for 3.5G/4G Systems

The deployment of 2.5G and 3G networks is continuously on the rise due to capacity constraints and wireless data demands. One could argue that this extra supply of network capacity could soon run out as a result of much higher demand. To keep up with the supply and to enable new data applications, operators are looking into technologies that could be integrated into their existing 2.5G/3G networks.

Similarly, one clear weakness of the existing 3G deployments is that as the number of voice users increases, the data throughput decreases. The decrease might not be noticeable if there are only few data users, but as the number of data subscribers increases the operators are no longer able to honor their commitment to the advertised data rates. Thus, the operators are constantly looking for ways to keep their promises.

In this section, we discuss techniques that will enable the operators to keep their commitments. The techniques, such as receive/transmit diversity, interference cancellation, and smart antennas, are available for 3G networks except for single antenna interference cancellation (SAIC), which is only applicable to GSM, but it is a highly desirable feature as we will see later in the chapter. The techniques such as MIMO, scalable bandwidth, and IP RAN are more applicable to 3.5G and 4G systems.

12.6.1 Mobile Receive and Transmit Diversity

Antenna diversity is one of the most popular forms of diversity used in wireless communication systems. In mobile receive diversity (MRD), multiple, spatially separated antennas are used to receive uncorrelated signals. Conventional cellular radio systems consist of an elevated BS antenna and a mobile antenna close to the ground. The existence of a direct path between the transmitter and the receiver is not guaranteed, and the possibility of a number of multipaths in the vicinity of the mobile device suggests a Rayleigh fading signal. The idea behind the MRD approach is to combine signals from two separate antennas in the handset to reduce the impact of spatial variations in signal strength and thus increase the average data rate and voice capacity. In addition to two antennas, MRD uses two receive RF chains to improve signal reception in a mobile phone (see Figure 12.1). One antenna provides both transmitting and receiving functions, while the second antenna can *only* receive signals. The distance between the antennas could be as little as 1 inch [8].

The MRD can double the number of voice users and the data capacity, but the addition of elements such as a second antenna and RF components will be costly.

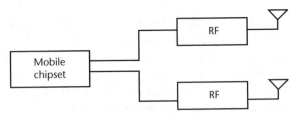

Figure 12.1 Mobile terminal with diversity [8].

The MRD-based solutions will be expensive to deploy beyond high-end devices and data cards.

The mobile transmit diversity (MTD) scheme adds a transmitting antenna in the handset for better communications with the base stations. Currently, MTD research is under way and the improvements have not received a sign-off from many players, although it expects to provide gains similar to those of MRD.

12.6.2   Single Antenna Interference Cancellation

SAIC is applicable to GSM-based networks and can not only improve the voice quality but also increase the voice capacity. It cancels the interference in the modulated signal (GMSK), thus allowing GSM operators to use a more efficient frequency reuse scheme that increases the voice capacity. SAIC only works with GMSK and is not applicable to the other higher modulation schemes (8-PSK, 16-QAM) associated with EDGE, HSDPA, and so forth. As a result, SAIC is not applicable to 3G networks; however, it can increase the capacity of GSM networks by at least 40% and improve the voice quality by 0.5 MOS (mean opinion score). With all such benefits, SAIC is now officially part of the 3GPP standard and a highly desired feature in GSM handsets for operators [9].

12.6.3   Interference Cancellation Technologies

The idea of interference cancellation (IC) is to estimate the multiple access and multipath induced interference and then subtract or project out the estimated interference. There are two forms of IC: successive IC (SIC) and parallel IC (PIC). The SIC receiver cancels estimated interference on a user-by-user basis, whereas the PIC receiver cancels the interference of all users simultaneously. PIC can also employ an iterative approach (multiple stages) if desired, but it would increase the complexity and cost of the underlying technique. On the other hand, SIC becomes unreliable for CDMA systems when signals from different users all arrive at the base station at the same time [10].

CDMA is the inherent access scheme of all of the 3G systems including WCDMA/HSDPA, TD-SCDMA, and CDMA2000/EV-DO and we all know that the capacity in the CDMA-based systems is interference limited. To minimize the impacts of interference, operators around the world are evaluating a handful of IC techniques.

The technique that is used to minimize the impact of the pilot channel is sometimes referred as pilot interference cancellation (PIC), whereas the one that reduces the impact of voice and data traffic is called traffic interference cancellation (TIC).

We can easily envision that TIC is much harder to implement than PIC, but if it gets implemented in the correct manner it could have a dramatic impact on reducing interference (and increasing capacity).

There is another interference cancellation scheme [11] that cancels interference from both pilot and traffic called the interference cancellation technique (ICT). Testing has shown an average increase of 2.5 dB in capacity for CDMA2000 systems. ICT adapts at the symbol rate, or the rate at which the bits are encoded/decoded as compared to equalizers, which adapt at the Doppler rate (low rate). Thus, ICT is better than equalizers because it can adapt to rapid changes in the characteristics of the radio link. Nevertheless, ICT requires integration into the baseband chipsets, which require a new spin to the chipset. Thus, it seems as if there will be a long lead time before the mass introduction of ICT into commercially available handsets.

As far as standardization activities are concerned, a study item has been introduced within 3GPP to investigate various IC technologies with the end goal of tighter requirements for WCDMA/HSDPA receivers.

12.6.4   Smart Antennas

Smart antennas have basically introduced the concept of adding antennas at the base station where each antenna is a few centimeters away from the closest antenna. By creating an antenna array, a radio signal from a subscriber arrives at each antenna with a different phase and time of arrival. Then DSPs are used to combine the individual signals into a single stronger signal. A smart antenna uses signal processing algorithms to identify the direction of arrival of the signal and to calculate beam-forming vectors that will track and locate the antenna beam on the mobile. Beam forming is used to create the radiation pattern of the antenna array by constructively adding the phases of the signals in the direction of the mobiles desired, and nulling the pattern of the mobiles that are undesired/interfering targets.

Smart antennas could be either fixed-beam or adaptive-beam antennas. There will be no movement of beams in the fixed-beam solution, and the gains will be lower than that of the adaptive-beam solution. The adaptive antennas are designed to combat the impacts of ISI and CCI. As ISI and CCI increase, the conventional (array) antennas are no longer effective. As the user moves within the cell/sector, the arrays from the adaptive antennas at the base station follow the subscriber to give him or her the largest amount of energy with the least amount of power. Thus by proper allocation of signal and power, a more than 2-dB capacity improvement can be achieved.

To a reasonable extent, smart antennas are a bit late to the 2G/2.5G game to solve coverage and capacity issues. Operators already have the mechanisms in place to address these concerns with cell splitting, additional spectrum, additional carriers, and so forth. Smart antennas are more likely to be integrated into 3G/3.5G (CDMA2000, HSDPA, 3G-LTE, and so forth) networks.

12.6.5   Scalable Bandwidth (Multicarrier CDMA)

The scalable bandwidth or multicarrier CDMA is a method to amalgamate two or more carriers. The combination of carriers as seen in the case of EV-DO Revisions B/C will increase data rates and data throughputs. Although, EV-DO Revision B can support 20-MHz channels, the commercial deployments are only expected to support three 1.25-MHz carriers in the 5-MHz bandwidth. On the other hand, EV-DO Revision C systems are expected to support up to fifteen 1.25-MHz channels in the 20-MHz bandwidth. In addition, 2.5-, 10-, and 15-MHz channels are also available. It is not necessary to use the same number of carriers in both the forward and reverse links; that is, an operator can deploy three concatenated carriers in the downlink and two in the uplink or vice versa.

12.6.6   Multiple Input Multiple Output

MIMO is one of the key technologies that has been strongly considered for future mobile networks. MIMO is explored in the downlink because most of the commercial 3G/3.5G systems are downlink limited. MIMO technology exploits multiple antennas both at the base station transmitter and user device receivers in the downlink.

In the base station transmitter, the data channel intended for the user is encoded in time and space across multiple transmit antennas. The process is conducted with the same carrier at each transmit antenna. Signal processing is then used to decode composite signals at the user's end where the spatial antenna processing is able to resolve multipath scattering. In this way, MIMO provides multiple parallel pipes between the base station and the user as compared to the existing scenario where only a single data pipe is available between the two entities. Thus, MIMO is able to capitalize on the presence of multiple pipes improving both the data performance and system capacity.

MIMO comes in various forms including 2x2 MIMO, 2x4 MIMO, and 4x4 MIMO.[3] Figure 12.2 shows a conceptual view of a 4x4 MIMO scheme, which consists of the RF front end, baseband front end, and MIMO decoder. As the number of transmit paths increases, the data performance also improves with 4x4 MIMO and can provide 3.0-Mbps data sector throughput in 3.5G systems.

12.6.7   Vocoders

Though the focus of futuristic wireless is on high-speed data, it is important that we briefly look into the evolution of speech coders for both circuit-switched voice

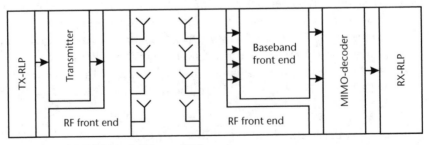

Figure 12.2   A 4x4 MIMO block diagram [12].

---

3.   4x2 MIMO has four antennas at the base station and two at the user's handset.

(CSV) and packet-switched voice (PSV). Adaptive Multi-Rate (AMR) and Enhanced Variable Rate Codec (EVRC) are the popular speech codecs that are used by the GSM/UMTS and CDMA2000 systems, respectively. AMR has eight voice sampling bit rates, whereas EVRC has three to support acceptable frame erasure rates (FERs). AMR wideband (AMR-WB) and AMR-WB+ vocoders are also being looked at by the GSM community.

The next wave of codecs will use multiple modes with each having a different data rate according to the network conditions. Selectable mode vocoders (SMVs) were one of the first to introduce the concept of modes; it was technically sound but it failed to prove its business justification for many operators. One of the recent vocoders is 4GV, introduced by Qualcomm. 4GV is a core voice codec suite that supports both CSV and PSD (i.e., CDMA2000 1X and EV-DO Revision A, respectively). This vocoder was specifically designed to leverage 1xEV-DO Revision A and to support delay-sensitive applications such as VoIP while maintaining the overall quality of service.

12.6.8   Flat Architecture

These days there is a push in the industry to consolidate the network elements associated with the access network. There are potentially two phases to this approach. The first is to consolidate some of the BSC/RNC functionalities into the BTS/Node-B and move the rest into PDSN/GGSN cabinets. In the final phase, the three network elements (BTS, RNC, and PDSN) are combined into one single element. We expect some level of consolidation in the later part of this decade.

12.7   Smart Phones

Smart phones are commonly defined as convergence devices that include cellular telephones, programmable information management features, and Internet access. With one such handheld device, one can access corporate e-mails, personal e-mails, text messages, the Internet, and regular cellular phone functions such as the phone book, internal storage (memory), and the calendar. Smart phones also allow users to play 3D games, listen to MP3s and FM radio, and record and play back videos. Many recent smart devices offer Bluetooth connectivity, which wirelessly connects to nearby devices, such as headsets, PCs for synchronization, and other smart phones.

Smart phones are available in multiple forms from many handset manufacturers and can operate in almost any cellular network. The most widely used smart phones are based on Symbian operating systems (OS). The Symbian platform has the support of many key players including Nokia, Sony/Ericsson, and Samsung. Samsung SGH-D730 and Sony Ericsson P910 are some of the more popular Symbian-based devices. The other key player is Microsoft, which supports three separate mobile OSs, all based on windows CE. Last but not least is the Palm OS, which has the broadest base of applications support. Treo 650 from PalmOne supports many smart phone features plus it provides connections for both GSM and CDMA. That said, issues remain, such as vendor support for multiple OSs, security concerns and

WiFi support. The 4G smarter devices are expected to be better than today's smart phones because they will have more processing power, better battery life, and more applications [13].

12.8 4G

Does anyone know what is meant by 4G? Everyone has his or her definition. We present a high-level action plan in this section to consolidate the four individual evolution paths of 3GPP, 3GPP2, CCSA, and IEEE that could lead to *one single mobile wireless communications standard* (4G) as originally dreamed of by ITU for previous generation systems (Figure 12.3). The 4G standard will primarily need to define a mechanism based on the OSI protocol stack so that it can provide a baseline document for the telecommunications industry. The manufacturers and operators could then add their specific functionalities on top of this protocol stack to attract operators and subscribers, respectively. The set of applications (discussed in the next chapter) will be the key differentiation points among operators rather than the underlying technology. Thus, defining multiple standards will not be so effective and will not generate economies of scale.

The 4G standard should provide a smooth migration for the 3G/3.5G systems and make the transition as painless as possible. The telecommunications industry has invested billions of dollars in 2G/3G networks and without a good return on investment, the operators and manufacturers will not make a serious move toward 4G. The convergence will expedite the commercial availability of the products, reduce costs, and offer seamless connectivity around the globe for some basic services such as voice, VoIP, news, game scores, and stocks update. For additional high-speed data services, extensive business agreements will need to be signed among operators.

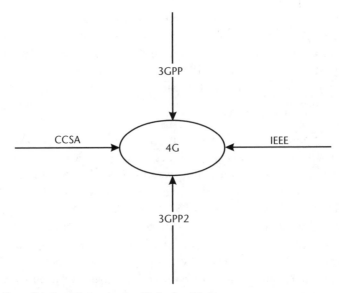

Figure 12.3 Consolidation of standards activity for 4G.

12.8.1 Convergence

The main task of 4G or convergence should be to bind the operators' WANs with the LANs (i.e., providing the last mile access). There are four possible choices from the global perspective: unlicensed mobile access (UMA), indoor fixed WiMAX and outdoor cellular, indoor WiFi and outdoor mobile WiMAX, and low-cost extension of the cellular network into the home or office. We envision that UMA, which is a standards-based approach defined for GSM/GPRS to WiFi handovers, will not be appropriate for 4G. The second alternative is to use indoors a fixed/nomadic WiMAX and a cellular network outdoors. This will allow wireless DSL via WiMAX in homes and businesses and provide mobility with a cellular network. The second alternative will have to resolve a number of WiMAX/cellular interoperability issues before it can become a reality. The third alternative, the WiMAX/WiFi hybrid, will need time to make WiMAX a true alternative to cellular in terms of mobility. The cellular industry has spend decades making GSM and CDMA technologies work in the mobile environment; future WiMAX operators need to do same.

The last solution, which we like the most, is to bypass WiFi and WiMAX all together and extend the wireless coverage into the home or office by means of very-low-cost picocells or femtocells. This solution will provide both indoor and outdoor coverage with one network (cellular) in contrast to the amalgamation of the two. This approach will provide a wider selection of handsets, which will be beneficial for consumers and businesses, but what is required is the availability of picocells and femtocells at a very low price. For 4G to succeed using this approach, it will have to overcome wired DSL and WLL solutions—and for that we not only need cheaper handheld devices, but also require a very high penetration of laptops, which is not currently feasible in developing countries.

12.8.2 Action Plan

We envision that technologies such as WiBro (broadcast wireless TV), WLAN, WPAN, WLL, and LMDS will become accessories to the 4G standard. The 4G standard will provide hooks to hand over control to these accessories as required. For example, the 3.5G/4G technologies will provide high-speed data cards that can be used in most parts of the city and also in hot spots and homes, making WLAN an unattractive choice. Also, some of the key features of Bluetooth and its successor UWB can be incorporated into 4G.

The current activities of the EV-DO Revision C, 3G-LTE, and 802.20 standards are scheduled to be baselined in 2007. The 802.16e (profiles) were finalized in 2006. The deployments of these technologies are not expected to start before 2009–2010 (Figure 12.4). However, ITU must start the work very soon, before further disintegration happens, and try to complete the standard by 2010. The 4G-enabled networks and devices could then expect to be available by 2012. To fulfill this task, ITU can pursue the following steps to make it a reality:

- Form a single committee comprised of members from 3GPP, CCSA, 3GPP2, and IEEE: 2007;
- Define roles and responsibilities: 2007;

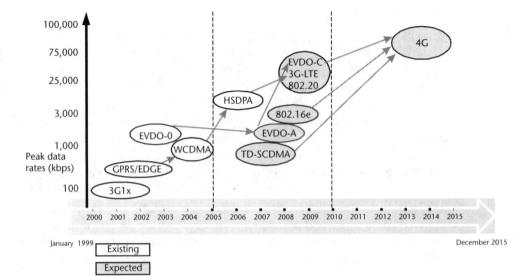

Figure 12.4 Evolution to 4G.

- Define key requirements for 4G networks and devices: 2007;
- Allocate frequency spectrum: 2008;
- Initiate proposals activity: 2008;
- Reach agreement on a consolidated proposal: 2009;
- Complete 4G specifications: 2010.

It normally takes 2 to 2.5 years after the completion of a standard to get the commercial products to market. The goal should be to synergize activities for 4G rather than bifurcating efforts as is happening with the 3G/3.5G standards.

12.9 Summary

In this chapter, we briefly defined the developments that are in progress for 3.5G technologies (EV-DO Revision C, 3G LTE, 802.20). The key technologies that would make 3.5G/4G technologies more attractive were also discussed. We also discussed smart phones and their applications and presented a high-level action plan for the consolidation of the different standards for 4G that would pave the way for *one single mobile wireless communications standard*.

References

[1]   CDMA Development Group; available at http://www.cdg.org.
[2]   GSM Association; available at http://www.gsmworld.com.
[3]   Thelander, M., "3GSM: ¿Dónde está...?," *Signals Ahead*, Vol. 3, No. 4, February 20, 2006.
[4]   "Updated Lucent-Nortel-Samsung Proposal for Air Interface Evolution Phase 2," 3GPP2 C30-20060522-032, 3GPP2 (Lucent, Nortel, Samsung), May 22, 2006.

[5]   "UHDR Overview," 3GPP2 C30-20060522-037, 3GPP2 (Qualcomm), May 22, 2006.

[6]   Thelander, M., "Welcome to the Evolution ... of 3G," *Signals Ahead*, Vol. 2, No. 26, December 19, 2005.

[7]   "MBFDD and MBTDD: Proposed Draft Air Interface Specification," IEEE C802.20-06/04, IEEE, January 2006.

[8]   Asif, S., "Mobile Receive Diversity Technology Improves 3G Systems Capacity," *Proc. IEEE Radio and Wireless Conf.*, Atlanta, GA, September 19–22, 2004, pp. 371–374.

[9]   Thelander, M., "Squeezing Blood from a Turnip," *Signals Ahead*, Vol. 3, No. 3, February 7, 2006.

[10] Ojanpera, T., and R. Prasad, *Wideband CDMA for Third Generation Mobile Communications*, Norwood, MA: Artech House, 1998.

[11] Asif, S., "Interference Cancellation Technique for CDMA2000 Handsets," *Proc. IEEE Vehicular Technology Conf.*, Dallas, TX, September 25–28, 2005, pp. 737–740.

[12] Adjoudani, A., et al., "Prototype Experience for MIMO BLAST for Third-Generation Wireless System," *IEEE J. on Selected Areas in Communications*, Vol. 21, No. 3, April 2003, pp. 440–451.

[13] Thelander, M., "When Mobile Viruses Attack (Film at 11:00)," *Signals Ahead*, Vol. 2, No. 16, August 15, 2005.

CHAPTER 13

Markets and Applications

13.1 Introduction

The mobile phone is becoming the "third screen" for users, and day by day they demand more from it. The first two screens, with which we are all familiar, are our TV and computer screens. Users are demanding very much the same level of experience from their cell phones as they are getting from their TVs and computers. We all know that customers care less about the underlying technology (i.e., GSM, CDMA, and OFDM) and more about the applications and services they can get from their respective service providers. As the wireless technologies mature and become more reliable, businesses are getting more and more interested in replacing their fixed, wired networks with wireless ones.

Thus, we have divided this chapter into two main segments. First, we discuss the applications in which the two different segments of the market—businesses and consumers—are interested. Although most of the applications are applicable to both segments, a few are exclusively used by one segment or the other. We also briefly address some of the key applications in this chapter. Second, an update of the wireless telecommunications industry for developed, emerging, and developing countries is provided.

13.2 Market Segments

The two segments of the market are businesses and consumers; their requirements and their appetite for data services are slightly different (Figure 13.1). Both segments desire the "triple play" (voice, video and data) to be available 24/7. Most of the applications that are highlighted in this chapter can be used by both segments, although the uses of some of these applications may be different among the two segments. Most of the applications are available today, but their performance is expected to improve with 3.5G/4G technologies.

13.2.1 Business Segment

The telecommunications needs for most of the businesses around the world are currently served by fixed, wired networks. To replace their wired infrastructures with wireless, businesses want at least a similar if not better experience at a reduced price for their ever moving mobile workforce.

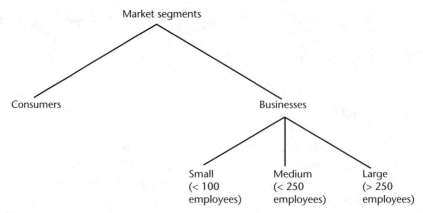

Figure 13.1 Market segments.

The business segment can be further divided into three main categories: small, medium, and large. The definitions of these categories in different parts of world vary, but a small enterprise is comprised of less than 100 employees, a medium-size enterprise consists of more than 100 but fewer than 250 employees, and large enterprises will have more than 250 employees.

The potential wireless alternatives for replacing corporate LANs are in-building repeaters, cheaper in-building micro/femto base stations, WiFi, and others. Any of these alternatives can be used to make the cellular tower signal more powerful within the building with the use of data connection cards in laptops (depending on the alternative).

Some of the key applications that businesses would like to have wireless alternatives for are discussed next.

13.2.1.1   E-Mail

Electronic mail is the most common form of information sharing within the same business and with other businesses. The term *e-mail* applies both to the Internet e-mail system and to intranet systems that allow users within one company to send messages to each other. Microsoft Outlook is one of the most popular personal information managers for e-mails.

13.2.1.2   Intranet

An intranet is used for providing access to a company's resources including human resources, travel and expense accounts, sensitive documents, news, and organizational structures. The access provided by an intranet enables employees to get the most up-to-date information available about their company.

13.2.1.3   Videoconferencing

A videoconferencing session allows two or more locations to interact simultaneously via bidirectional audio and video transmissions. Videoconferencing provides meeting attendees with the same feeling they would get at an in-person meeting. It also saves travel time and travel-related expenditures.

13.2.1.4 IPTV

Internet Protocol Television (IPTV) is a system in which digital television service is delivered to subscribers using IP over a broadband connection. It uses the Internet and sends less information than the standard analog or digital television, thus offering a lower cost solution. Another advantage of IPTV includes a two-way capability not provided by traditional TV systems that can be used by the CEO for Webcasts. IPTV is gaining significant momentum in terms of testing and deployment.

13.2.1.5 Wireless Priority Service

Wireless Priority Service (WPS) is a U.S. government initiative that will allow authorized National Security and Emergency Preparedness (NS/EP) personnel priority access to available cellular resources during emergency situations. WPS will not preempt calls in progress and does not guarantee call completion. Although WPS is a voluntary service, all the major U.S. operators are either already providing or will launch this service in 2006–2007. The details can be found at http://wps.ncs.gov/.

13.2.1.6 PTT

Push-to-talk (PTT) is a two-way communication service that allows users to engage in immediate communication with one or more users. PTT is a VoIP packet data call, which is different than the traditional circuit-switched calls. PTT service is similar to a "walkie-talkie" application where a user presses a button to talk with any individual user or to broadcast to a group of participants. The receiving participant hears the sender's voice either without any action on his or her part (for example, without having to answer the call) or may be notified and have to accept the call before he or she can hear the sender's voice. Other participants can respond to this message once the initial speech is completed. The communication is half-duplex; that is, only one person can talk at a time, and all other participants hear the speech. This contrasts with voice calls, which are full duplex, where more than one person can talk at the same time.

PTT is one of the popular applications for the business world especially in the United States. Construction and factory workers are the key recipients of this service. There are millions of PTT service subscribers in the world, and the number is expected to grow.

13.2.1.7 Mobile Medical Support

The true mobile broadband system will enable doctors and other medical practitioners to treat their patients on the go. They can also assist nurses or other medical personnel who may be present at the place where a medical emergency is happening. This application will come in handy during natural disasters and in places where adequate medical services are not available.

13.2.1.8   Surveillance

Surveillance is a hot topic these days, especially when the world around us is getting more and more insecure day by day. Currently, some 2G/3G-based wireless monitoring systems are available that instantly allow users to conduct monitoring anywhere on the Earth without being limited by the absence of the Internet. When one can witness any unnecessary presence in one's home from the mobile device, the application becomes really powerful. Even a Webcam can be used in the house for monitoring purposes.

13.2.1.9   Secure Cellular Network for Ground Forces

Another application considered by the telecommunications industry is one that will provide a small, secure cellular network for the military, especially when the usual means of communications are disrupted. It might be comprised of just two or three cell sites each a few kilometers long that could be established on the fly. It would provide encrypted voice and data communications within and between the sites and headquarters.

13.2.1.10   Electronic Commerce

E-commerce consists primarily of distributing, buying, selling, and servicing of products and services over the Internet. It also involves e-fund transfers, online banking, supply chain management, among others. A mobile phone can now offer more functions than just voice and Internet access. It can become one's wallet, credit card, ID card, or a key to one's home. Japan, as always, has been instrumental in enabling new services such as mobile commerce. In Japan, NTT DoCoMo was the first company to introduce handsets that can be used to make cashless payments in stores via a contactless interface developed by Sony. DoCoMo launched i-mode FeliCa wallet phone service in 2005. The service was enhanced in 2006 with the launch of the DCMX credit card phone. At the end of last year, just 5 months after the service was launched, the company reported 1 million of the handsets sold, and the number is expected to increase in the coming years.

13.2.1.11   Cell Phones on Airplanes

A number of operators have tested the capabilities of using cell phones during airplane flights. Tests have shown no interference between radio communications. The regulatory bodies are still looking for more data from the operators before they allow voice and data communications via cell phones during airplane flights.

13.2.1.12   Wireless Backhaul

The advent of wireless broadband is forcing operators to look for alternatives that can replace their expensive T1/E1 backhaul lines.[1] The deployment of EV-DO

---

1.   T1 (Transmission Level 1): 1.54 Mbps in North America and Japan; E1: 2.048 Mbps outside North America and Japan.

Revision A will require operators to deploy three or four T1 lines per tower instead of just one with 3G1x. This number will definitely rise with 3.5G/4G technologies. The alternatives are available in the form of point-to-point systems such as WiMAX, LMDS, and FSO.

Figure 13.2 shows the potential penetration of the business-oriented applications in 2007 for the three different types of markets.

13.2.2   Consumers Segment

Today, the most appealing and most demanding application for consumers is circuit-switched voice. It is expected to be replaced by VoIP (packet-sized voice) in the next few years. The emergence of data connection cards and increased use of laptops is pushing the envelope on the consumer side. The most common method of acquiring broadband home infotainment is fixed DSL. In the coming years DSL is expected to get fierce competition from wireless broadband technologies. Some of the key applications for this segment are discussed next.

13.2.2.1   Voice

As voice revenues decrease, the industry is looking to replace circuit-switched voice with VoIP. VoIP will further reduce calling charges for consumers since it will be cheaper for operators than the traditional voice. Mass market adoption of VoIP

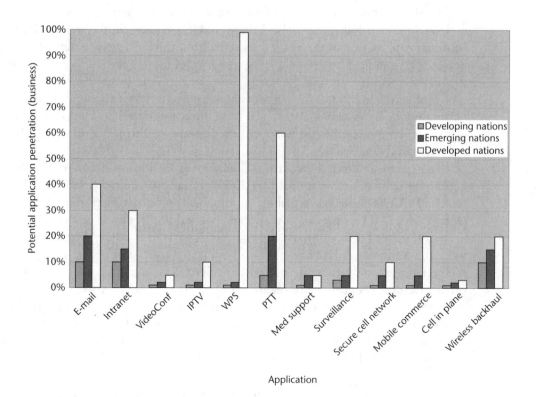

Figure 13.2   Potential business-oriented application penetration in 2007.

over mobile networks is expected in the next 2 to 5 years with the deployment of 3.5G technologies. Most of the 3G technologies are not suitable for VoIP because of their high overhead, which could significantly reduce system capacity.

13.2.2.2 Personalized Communications

Personalized communications consist of applications and services that are based on subscriber's personal data. These could include calendar/schedule, messaging, phone book, messaging, chat, and ring tones.

13.2.2.3 Digital Audio Broadcasting

Digital audio broadcasting (DAB) is a technology for broadcasting audio programming in digital form that was designed in the late 1980s. The standardization of DAB technology is promoted by the World DAB Forum, which represents more than 30 countries excluding the United States. From a radio listener's perspective, the difference between DAB and FM radio is primarily the availability of the number of stations and the audio quality. DAB (digital) can carry many audio services on a single frequency, whereas FM (analog) can only carry one per frequency. DAB has been introduced in many countries around the world.

13.2.2.4 Mobile TV

Today, the most common distribution path for multimedia services is through cable and satellite distribution models. As new services emerge in the world of wireless communications, one of the challenges is the broadcast of video content to a mobile device. The mobile TV market is getting more and more upbeat every day. Some of the major operators around the world are offering some form of mobile TV. If we only look at the market in the United States, consumers spent around $80 billion in nonmobile video services in 2004. The good thing about mobile TV is that the traffic is mainly unidirectional (downlink), which saves quite a bit of spectrum and resources in the uplink.

There are five likely choices for the delivery of TV-like service to mobile devices: digital video broadcast–handheld (DVB-H), broadcast multicast service (BCMCS) for CDMA2000-based networks, multimedia broadcast and multicast service (MBMS) for UMTS systems, DAB-IP, and MediaFLO (forward link only). The primary difference between broadcast and multicast is as follows: When broadcast is used, the transmitted data are sent to all interested users within a given region with the system assuming that the user(s) is present. The signal is also transmitted or broadcast into every cell sector regardless of whether or not there is a targeted recipient in the area. With multicast, an additional layer of intelligence is added in that the network resources are only used if the subscriber has registered for the service and he or she is present to receive it. This additional functionality saves unnecessary use of radio resources [1].

- The DVB-H standard has been developed for the delivery of audio and video content to mobile handheld devices. It overcomes two key limitations of the earlier version of the standard: It lowers battery power consumption and improves robustness in poor RF conditions. DVB-H can be used along with cellular technologies (CDMA2000 and UMTS) and thus benefits from access to a cellular network as well as a broadcast network.

- BCMCS offers CDMA2000-based networks an opportunity to go beyond the concept of unicast to enhance the customer experience by broadcasting TV-like audio and video services. Platinum Multicast (a form of BCMCS) is expected to be available with EV-DO Revision A in the coming years, whereas Gold Multicast services are available today with EV-DO Revision 0.

- MBMS is part of 3GPP Release 6 and is similar to other broadcast technologies. It is important to note that the uplink frequency bandwidth will be not fully utilized if MBMS is offered via WCDMA-FDD–based systems. Although MBMS brings important broadcasting enhancements to WCDMA, it cannot match the capabilities of DVB-H.

- The DAB standard has been revised by ETSI to accommodate transmission of mobile TV via IP (DAB-IP). DAB-IP is more suitable than DAB for mobile radio and TV because it has better voice codecs compared to the original DAB. DAB-IP has more potential than MBMS to compete with DVB-H.

- Last but not least we have MediaFLO, a proprietary system developed by Qualcomm to deliver broadcast services to handheld receivers using the 700-MHz spectrum.

13.2.2.5 Internet Surfing

Another key application heavily used today is *Web surfing*. Web surfing is used to access news, sports, and weather information and also for file sharing. The Internet has also become one of the most popular forms of conducting research. *Music and movie downloads* are very popular among teens. The Internet allows computer users to connect to other computers and access information, wherever they may be across the world (i.e., remote access). *Streaming voice/audio* is another application that allows radio and television broadcasters to provide Internet "feeds" of their live audio and video streams. *Video-on-demand* systems allow subscribers to select and watch video content over a network as part of the interactive television system. Most of a DVD player's features, such as fast forward and pause, also work. Current mobile devices do not support broadband Internet services. Future 3.5G/4G systems are expected to compete to replace the existing DSL services.

13.2.2.6 Messaging

The Short Messaging Service (SMS) is one of the most popular applications used by the consumers. SMS permits the sending of short messages (text messages) between mobile phones, other handheld devices, and even landline telephones. It saves money and time. Approximately 235 billion SMSs were sent during just the first quarter of 2006 [2].

13.2.2.7   Interactive Gaming

Interactive gaming is another area that has been introduced by several operators around the world. The current 3G networks are not most suitable form for interactive gaming. The deployment of new technologies as discussed in the previous chapter will make the process more attractive.

13.2.2.8   Sharing

Most of today's mobile phones have built-in cameras. The ability to take pictures has become increasingly common and popular in mobile communications. This functionality allows the user to share pictures with friends and family in real time. It has also been used to capture thieves. Mobile Picture technology from Microsoft defines a set of features to aid common image browsing tasks, such as viewing thumbnails, zooming, and scrolling based on image features to further enrich the experience.

13.2.2.9   Location-Based Services

Location-based service (LBS) applications are provided to users based on their current geographical location. The location is usually determined by a user entry or a GPS receiver that he or she carries along. One example of LBS is that of finding a nearby move theater or restaurant.

Figure 13.3 shows the potential penetration of the consumer-oriented applications in 2007 for the three different types of markets.

13.3   The Mobile Telecommunications World

Today's mobile telecommunications world is changing at a pace that was unpredictable 5 years ago. Most of the growth is happening in the emerging and developing nations, whereas most of the R&D activities are taking place in the developed countries. As shown in Figure 13.4, countries such as the United Kingdom, Japan, and the United States are considered developed; China, India, and Malaysia are considered emerging markets, while Pakistan, the United Arab Emirates, and Morocco are considered to be developing nations. This section briefly reviews the current position and the future outlook for these three major telecommunications markets.

13.3.1   Developing Nations

Mobile telecommunications markets in the developing nations are on the move. In the past 5 years, these markets have witnessed a number of changes and gone through a transformation cycle. Some of the key reasons for this transformation are as follows:

- *Telecommunications market liberalization (TML):* One of the most important factors for this transformation is the liberalization of the telecommunications market. TML was necessary for countries to gain World Trade Organization

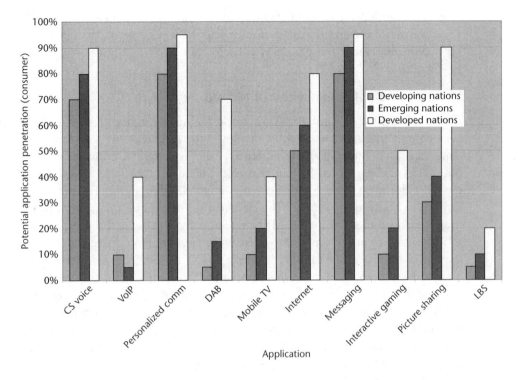

Figure 13.3 Potential consumer-oriented application penetration in 2007.

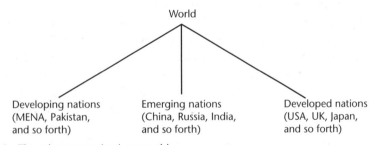

Figure 13.4 The telecommunications world.

membership. TML is on the rise in developing nations and involves formation of a regulatory body (independent from the Ministry of Communications). This body implements laws via deregulation, issuance of new licenses, and monitoring of privatization activities in the country.

- *Competition:* As new operators enter the market, the incumbent(s) have to protect their existing base, which results in price reductions for customers. Sometimes the new entrants are big regional or global operators, making the tasks for the incumbents more difficult.

- *Technology:* Another important factor is the technological component in this race. Although voice communications remain the bread and butter for the operators, its average revenue per user (ARPU) keeps decreasing. Thus in the short to long term, the operators have to add data and other value-added services (VAS) to the equation. In the coming years, data and VAS will become the major differentiation factor among the service providers desiring to retain old subscribers and gain new subscribers.

13.3.1.1 Arab Telecommunications Markets

During the past few years, the Arab world has seen a major change in the telecommunications market. The Middle East and North Africa (MENA) includes 20-plus countries. The countries in alphabetical order are Algeria, Bahrain, Comoros, Djibouti, Egypt, Iran, Iraq, Jordan, Kuwait, Lebanon, Libya, Mauritania, Morocco, Oman, Palestine, Qatar, Saudi Arabia, Somalia, Sudan, Syria, Tunisia, United Arab Emirates (UAE), and Yemen. The penetration of mobile telecommunications has dramatically increased in the MENA region from 1% of the total population in 1997 to more than 20% in 2005. There were more than 35 operators in the region at the end of 2005 and the number is on the rise. The operators are offering at least 20 postpaid and prepaid plans to their customers. Jordan is by far the most competitive market; it has four licensed mobile operators (three GSM and one IDEN), whereas Palestine and Qatar only have one mobile operator. The governments of Saudi Arabia and the UAE have recently provided licenses to a second mobile operator. The operators are Etihad Etisalat in Saudi Arabia and Emirates Integrated Telecommunication Company PJSC (EITC) in UAE. They are expected to begin providing service in 2006–2007.

Another change that has been witnessed by the region is the *acquisition of licenses* by outside operators. There are three reasons for this new endeavor: saturation in the home market, competition from the new operators in the home market, and opportunities in neighboring (or even outside the MENA region) countries. Etisalat, MTC, and Orascom Telecom[2] have more than 7, 13, and 10 operations, respectively, throughout the world. Etisalat added several countries in Central Africa, MTC has more than 13 operations in the Africa, and Orascom Telecom has expanded its base from Egypt to Africa, Bangladesh, and Pakistan.

Another change that is about to happen on a larger scale is the advent of *broadband wireless and 3G* in the MENA region. As the voice ARPU decreases day by day and users are leaning more and more toward data-rich services, operators are looking to deploy 3G and broadband wireless technologies. A handful of nations have already deployed 3G and EDGE technologies and the number is increasing. MTC-Vodafone launched the WCDMA network in Bahrain (first in Middle East) in 2005. WiMAX has also entered the market and multiple trials are under way.

If we peek into the near future, several developments are expected to happen:

- Potential consolidation as larger operators see their revenues go down because of smaller operators;
- New deployments of 3G (WCDMA/HSDPA) networks;
- Possible deployments of pre-WiMAX and later 802.16e solutions;
- Convergence of media and telecommunications activities.

---

2. Etisalat, MTC, and Orascom are companies based in the United Arab Emirates, Kuwait, and Egypt, respectively.

13.3.1.2   Pakistani Telecommunications Market

The Pakistani telecommunications market was mended in 1996 with the passing of the Pakistan Telecommunications Reorganization Act of 1996. This act set the stage for market deregulation and established the regulatory framework. The Pakistan Telecommunications Authority (PTA) is the regulatory body of the country. The PTA is responsible for all the sectors of the telecommunications industry including fixed and mobile wireless, satellite services, Internet, paging, cable, and television services. The monopoly formerly enjoyed by Pakistan Telecommunications Company Limited (PCTL)[3] ended in December 2002, which triggered a wave of competition in both wired and wireless markets.

Cellular Market Overhaul

At the end of September 2002, there were approximately 1.5 million cell phone users in the country, representing roughly 1% of the total country population. To keep the spirit of liberalization on the roll, the government of Pakistan announced its mobile cellular policy in 2004 to attract two additional cellular players. Warid Telecommunications and Telenor Pakistan were granted licenses to operate GSM networks in the country. With these two additions, there are six major cellular providers in the country with Mobilink being the market leader. This addition of players and new cellular policy have led to an enormous growth in Pakistan's mobile cellular industry. At the end of the second quarter of 2006, there were more 31 million mobile users, which represents a more than 2000% increase in less than 4 years. Mobilink, also called Pakistan Mobile Company Limited (PMCL) is a subsidiary of Orascom Telecom; Etisalat owns 20% of PTCL including management control of the company (Table 13.1).

As with other markets, as the voice ARUPs decrease for all cellular providers in Pakistan, most of them are looking to invest in broadband wireless data. They are also investing heavily in VASs such as SMS, ring tones, IP telephony, and multimedia. The companies are exploring DVB-T/H, pre-WiMAX, and WiMAX solutions to bring revenues to their respective tables. In May 2006, Wateen Telecommunications

Table 13.1   Pakistan Cellular Companies

Mobile Operator	Approximate Number of Subscribers	Description
PMCL (Mobilink)	15+ million	GSM 900/1800; Orascom has 89% ownership plus management control
Ufone	5+ million	GSM 900; Etisalat has 100% management control
WaridTel	4+ million	GSM 900/1800; company is owned by the Abu Dhabi Consortium of the UAE
Telenor Pakistan	2+ million	GSM 900/1800; Norwegian incumbent operator Telenor has 100% ownership
Paktel	1 million	GSM 900; Millicom International Cellular S.A. (MIC) has 98.9% ownership
Pakcom (Instaphone)	0.5 million	TDMA 800; MIC (61.3%) and Arfeen Group (38.7%)

---

3.   PTCL was the incumbent service provider for all the telecommunications services in the country.

(the long-distance arm of WaridTel) made the announcement that it would deploy the first 802.16e system in the world. The PTA was also expected to auction 3G (UMTS) 2,100-MHz licenses in mid-2007 with possible deployments of W-CDMA/HSDPA in 2008–2009.

13.3.2 Emerging Economies

The most impressive emerging economy is, as we all know, China. Besides China, Brazil, India, Malaysia, and Russia have also seen a huge growth in telecommunications needs during the past 10 years. We provide brief highlights of the Chinese, Indian, and Russian telecommunications markets in this section.

13.3.2.1 China's Mobile Sector

If we look back in time, the Chinese government has maintained a policy of reform since the 1978 opening of China to rest of the world. This policy has benefited the Chinese telecommunications market, which has experienced rapid growth since that time. At the end of 1992, there were approximately 19 million telephone sets, representing a growth of 68 times compared to 1950. From the mid-1990s until now, the Chinese telecommunications sector has exploded, and China became the world's largest telecommunications market in 2002, ahead of the United States. The Chinese telecommunication industry's growth was about 20% between 1997 and 2002. It has the largest number of GSM users (>320 million) and it is the number one mobile country with more than 400 million mobile subscribers overall [2–5].

Like other developing nations, China joined the WTO on December 11, 2001, which resulted in the formation of a regulatory body and steady opening of telecommunications services market to foreign companies. Although China has joined the WTO, it does not have a telecommunications act in place and the regulations are promulgated by the Ministry of Information Industry (MII). Thus, the quickest and the best way to gain access to the Chinese market is for foreign investors to find a Chinese partner (preferably a major carrier) and form a joint venture.

There are two major mobile operators in China: China Mobile and China Unicom. The state has control of and majority ownership of both of them. China Mobile is China's largest mobile operator and it provides GSM/GPRS services. It also provides data services, VoIP, and Internet access. It ranks first in the world in terms of network scale and customer base with close to 250 million subscribers. China Unicom was established in 1994 and is the only Chinese operator with a license to provide a full range of telecommunications services including mobile. It launched GSM services in 1994 and in 2002 also launched CDMA services to become the nation's only operator with a CDMA network.

In the past, the main telecommunications manufacturers were foreign but now Chinese manufacturers such as Huawei Technologies and ZTE are leading this market as well. Two main reasons for their growth were government support for the growth of local telecommunications companies and the development of homegrown technologies, such as TD-SCDMA.

Hong Kong Market

The Office of Telecommunications Authority (OFTA) is the legislative body responsible for telecommunications regulations in Hong Kong. The OFTA has fully liberalized all telecommunications sectors and there are no foreign ownership restrictions. The OFTA awarded four 3G licenses in 2001 to the four major operators, while the first 3G mobile service was launched in January 2004. All together there are 11 GSM networks and the mobile density is one of the highest in the world (121.40% mobile penetration rate).

Coming Years

The mobile segment of the telecommunications industry is expected to get the largest share of the pie in terms of investment, but both China Mobile and China Unicom are also expected to increase spending on wireless broadband, VASs, VoIP, and 3G (TD-SCDMA). It was widely expected that the MII would award the first 3G license in 2006 and that TD-SCDMA would receive favorable treatment from the government. TD-SCDMA network coverage is expected to be nationwide, whereas WCDMA networks will probably be deployed in certain areas [1]. It is also expected that China will have more than 500 million mobile users by 2010.

13.3.2.2   India

During 2005 India as emerged as the sixth largest cellular market in the world in terms of its subscriber base. Currently there are more than 60 million mobile (GSM and CDMA) users in India. In the next 2 to 3 years the Indian market is expected to account for 8% to 12% of the global handset market and to have close to 100 million subscribers.

There are more than 10 mobile operators in the country with Reliance, Bharti, and Bharat Sanchar Nigam Limited (BSNL) sharing the majority of the subscribers. Both Reliance and BSNL support GSM and CDMA systems; Bharti has deployed a GSM/GPRS-based network.

WCDMA deployments are expected in mainly urban markets, such as Delhi and Mumbai, in 2006–2007. On the other hand, EV-DO deployments are only expected if the operators are allowed to deploy the technology in the 850-MHz spectrum since EV-DO systems are not available in the 2,100-MHz band. A number of Indian service providers are also looking to deploy WiMAX and extend broadband services to the masses [1].

13.3.2.3   Russia's Mobile Sector

Russia is world's largest country in terms of area—it is 1.8 times the size of the United States. Because it spans 11 time zones from Eastern Europe to Northern Asia, Russia's population of 144 million represents a huge telecommunications market with substantial potential for growth.

Prior to the early 1990s, the telecommunications network in the former Soviet Union was unreliable and underdeveloped. In the early 1990s, the Ministry of Communications relinquished operational control to a single long-distance and international carrier, Rostelecom, and 80 incumbent regional operators. The formerly state-owned local operator could not keep up with the modernization of telecommunications; as a result, the Ministry of Communications issued licenses to domestic and foreign-funded companies to encourage investment in the telecommunications infrastructure. A new law on communication, effective since January 1, 2004, allowed regulation of ownership, distribution of RF bands, and licensing [6].

The Russian telecommunications market environment is in constant evolution and is gradually moving toward a more structured environment. The consolidation of the 72 fixed local incumbents into seven regional operators under the state-controlled giant Svyazinvest and the consolidation of the mobile market under three key operators translated into more efficient, better quality, and lower priced services. In 2005, 42% of Russia's telecommunications revenue was generated by mobile services, compared with 26% in 2000. The fixed long-distance revenue has, however, dropped significantly to 12%, compared with 32% in 2000. This was mainly due to the reduction in tariffs and the growth of the mobile sector.

Mobile services continue to be the key driver of the Russian telecommunications market, representing 75.7% of all telecommunications connections in the country. In 2005, the mobile market attracted 51.4 million new subscribers, 88% of whom are outside Moscow. At the end of 2005, there were 125.8 million mobile customers, representing a growth of 69.0% in that year alone. Mobile ARPU is continuing its downward trend and mobile saturation is now on the rise; the operators are starting to look toward neighboring countries and VASs for growth. It is forecasted that the number of mobile subscribers will increase by an average of at least 9.0% in the next 5 years to reach approximately 190 million by 2010. 3G licenses were also expected to be issued in the second half of 2006 with the goal of having service launched in 2007.

13.3.3 Developed Nations

Countries such as the United States, most countries in Europe, Japan, and South Korea all have well-developed telecommunications infrastructures. Most of the R&D and standardization activities also take place in these countries. A number of nations have deployed 3G networks including WCDMA, HSDPA, 3G1x, and 1xEV-DO. In the following subsections, we primarily look into Japan and U.S. mobile markets.

13.3.3.1 Japan

In Japanese, mobile phones are called *keitai denwa*, which means "portable telephones," and they are often known simply as *keitai*. *Keitai* refers to Web access, e-mail, games, and so on, not just a voice-enabled handset. The particularities of their usage have led to the development of a mobile phone culture or *keitai* culture [5].

More than three-quarters of all Japanese use their cell phones to surf the Internet. Wireless data access fees are relatively low because providers charge flat rates. Japanese operators have launched many revenue-generating applications such as picture messaging (sha-mail), mobile Internet access (i-Mode), video-conferencing, and multimedia downloads (i-motion). Japanese mobile phones have the capability to use very large sets of characters and icons based on Japanese Industrial Standards (JIS) that define characters for industrial appliances. More than 1,000 characters, including all of the Latin alphabet, different Japanese writing systems, and special characters like cm (for centimeter), arrows, musical notes, and more can be used to compose messages. There is popular trend in Japan to use two-dimensional bar codes for advertising. SMS messaging on newer, thinner color phones is very popular among teenagers.

Japan has three main mobile service providers:

- *NTT DoCoMo:* The predominant operator in Japan with more than 50 million subscribers. DoCoMo was the first operator in the world to launch the 3G mobile service based on W-DMA in 2001. DoCoMo also introduced i-Mode, a proprietary mobile Internet platform that has a worldwide presence. DoCoMo has launched HSDPA and is conducting R&D for 3G-LTE.
- *KDDI:* The next in line with more than 23 million subscribers. KDDI operates a CDMA2000 (1x and EV-DO) network. In 2006–2007 KDDI is expected to launch an EV-DO Revision A network.
- *Vodafone KK Japan:* The third largest operator in Japan with more than 15 million users. Vodafone KK (to be changed to SoftBank Mobile Corp., effective October 1, 2006), also known as Vodafone Japan and previously as J-Phone, was the Japanese subsidiary company of mobile phone operator Vodafone. On March 17, 2006, Vodafone Group announced that it had agreed to sell Vodafone KK to SoftBank for approximately 1.75 trillion Japanese yen (approximately US $15.1 billion). It currently operates both 2G (PDC) and 3G (WCDMA) networks.

13.3.3.2   U.S. Mobile Market

The United States has four major mobile operators: Cingular, Verizon, Sprint Nextel, and T-Mobile. Both CDMAOne/CDMA2000 and GSM/UMTS systems have been deployed in the United States. At the end of 2005, there were more than 185 million wireless subscribers in the United States, representing roughly 65% of the total population.

- *Cingular Wireless:* The largest U.S. mobile phone company, and it operates a network of multiple technologies. It has deployed GSM/GPRS systems and also an upgrade for faster speeds called EDGE. Cingular has launched a high-speed network known as "BroadbandConnect," based on UMTS and HSDPA, to counter Verizon Wireless's and Sprint's EV-DO networks. Cingular had a customer base of more than 53 million people.
- *Verizon Wireless:* Operates the second-largest wireless telecommunications network in the United States. As of January 2006, the company served a total of 51.3 million customers but had the largest service by area. The company is a joint venture of Verizon Communications and Vodafone Group, with 55% and 45% ownership, respectively. It has deployed 3G1x and EV-DO Revision 0 systems and is planning to launch EV-DO Revision A in 2007–2008.

- *Sprint Nextel:* One of the largest telecommunications companies in the world, with a combined market capitalization of $77.1 billion as of early 2006. With 51 million wireless subscribers, under both the Sprint PCS and Nextel brands, it became the third largest wireless service provider in the United States. It has deployed CDMAOne and CDMA2000 1X nationwide, while EV-DO Revision 0 services are offered in selected markets. The company is planning to launch EV-DO Revision A service nationwide in 2007–2008 in its 1,900-MHz spectrum. Sprint also holds the largest share of the BRS spectrum (2.5 GHz) in the United States and has recently selected WiMAX technology for its next-generation network.

- *T-Mobile:* A multinational mobile phone operator. It is a subsidiary of Deutsche Telekom and belongs to the FreeMove alliance. Globally, T-Mobile has 120 million subscribers, making it the world's third largest mobile phone service provider, and the second largest multinational after the United Kingdom's Vodafone. T-Mobile exclusively uses the GSM 1,900-MHz frequency to build networks and has also deployed EDGE. T-Mobile USA is planning to bid for 3G spectrum (in the 2,100- and 1,700-MHz bands) in the upcoming FCC auctions, to be held in June 2006. The 3G network is expected to be based on a UMTS/HSDPA solution with a possible launch in 2007.

According to the 2005 *Telecommunications Market Review and Forecast,* an annual publication of the Telecommunications Industry Association, the U.S. wireless market will reach $212.5 billion by 2008, with a 10% compound annual growth rate from 2005 to 2008. The report also predicts that there will be 278.5 million wireless subscribers by then, representing 88% of the population [7].

13.4   Summary

In summary, we presented information about the two main segments of the wireless market: businesses and consumers. A number of applications were also discussed for these two segments. In the second part of the chapter telecommunications developments were discussed for developing, emerging, and developed nations. A lot of growth is expected in the mobile wireless markets of the MENA region, Pakistan, China, and India, while market saturation is expected in Europe and the United States in the coming years.

References

[1]   Thelander, M., "2006—The Year of the ...," *Signals Ahead,* Vol. 3, No. 1, January 09, 2006.

[2]   CellularOnline; available at http://www.cellular.co.za.

[3]   Yunqian, C., "Driving Forces Behind China's Explosive Telecommunications Growth," *IEEE Communications Magazine,* Vol. 31, No. 7, July 1993, pp. 20–22.

[4]   Wikipedia; available at http://en.wikipedia.org.

[5]   Paradis, I., and Yi, M., "China—The World's Largest Telecommunications Market and More to Come," *Hot Telecom*, March 2006, pp. 1–4.

[6]   Paradis, I., and M. Yi, "Russia—Full Steam Ahead with Liberalization," *Hot Telecom*, April 2006, pp. 1–3.

[7]   TIA, http://www.tiaonline.org/business/media/press_releases/legacy.cfm?parelease=05-05.

About the Author

Saad Zaman Asif has close to nine years of experience in evaluating state-of-the-art wireless technologies and engineering of fiber optics systems.

He began his career in 1998 as an engineer in Sprint Nextel's (formerly Sprint) Transmission Engineering group where he engineered DWDM systems. A year later he moved to Sprint Nextel's Technology Development (former Technology Research and Development) organization where he assessed and conducted proof-of-concept (POC) trials on a number of wireless technologies including 3G (CDMA2000), interference cancellation, and smart antennas. He also led teams in conducting POC tests of EV-DO technology and played a major role in designing Sprint Nextel's wireless high-speed data strategy.

He is currently working as a manager in the Technology Development group of Mobilink (an Orascom Telecom Company), Pakistan. His current focus is on broadband wireless technologies, including WiMAX and WiFi/WiMAX hybrid networks. He is also evaluating DVB and a number of value-added services.

Mr. Asif received a B.S. and an M.S. in electrical engineering from Oklahoma State University in 1996 and 1997, respectively. He also received an M.S. in engineering management from the University of Kansas in 2001. He has published several papers on wireline and wireless technologies. He holds a U.S. patent and has patent applications pending with the U.S. Patent Office. He is also a senior member of the IEEE.

Index

A

Adaptive multirate (AMR), 13, 251

Adjacent cell interference, 10

Advanced encryption scheme (AES), 160, 175, 223, 228, 232–233, 240

Advanced mobile phone system (AMPS), 7, 15–16

A-key, 220–221, 223

Arab telecom markets, 266

Authentication and key agreement (AKA), 220–221, 224–226, 239, 240

Average revenue per user (ARPU), 265–266, 270

B

Broadband wireless access (BWA)

broadband radio access network (BRAN), 149, 165

broadband radio systems (BRS), 153–165

free space optics (FSO), 166–169

HIPERACCESS, 165

HIPERMAN, 165–166

IEEE 802.16, 149, 153, 157, 160, 163–164

IEEE 802.16-2004, 157–159, 161, 163, 164,

IEEE 802.16e, 157, 163–164, 170

IEEE 802.20, 163–164

local multipoint distribution service (LMDS), 150–153

mesh network, 153, 155–161, 166

millimeter wave (MW, E-band), 169–170

mobile broadband wireless access (MBWA), 163

orthogonal frequency division multiplexing (OFDM), 153, 157–158, 161–164, 170

orthogonal frequency division multiplexing access (OFDMA), 153, 157–158, 162–164

point-to-multipoint (PMP), 150, 153, 155–156, 158, 160–161, 166

point-to-point (PTP), 155–156

scalable OFDMA (S-OFDMA), 163–164

worldwide interoperability for microwave access (WiMAX), 153, 155, 165

Broadcast multicast service (BCMCS), 68, 262–263

Business segment, 257

C

Code division multiple access-2000 (CDMA2000)

1xEV-DO, 42–64

1xEV-DV, 64–67

call processing, 38–39

capacity, 39–41

carrier spacing, 25

IS-2000 architecture, 81–82

IS-2000 enhancements, 67–69

IS-2000, 25

IS-856 architecture, 82–83

IS-856 enhancements, 69–70

IS-856, 25

layering structure, 27

link layer, 32–34

physical layer, 28–32

radio configuration, 25

radio transmission technology (RTT), 25

spreading rate, 25

upper layers, 34–38

CDMAOne (IS-95), 16

Cellular authentication and voice encryption (CAVE), 221–223

Cellular message encryption algorithm (CMEA), 222–223

Change of address (COA), 200

China Mobile, 268–269

China Unicom, 268–269

China's mobile sector, 268–269

Cingular Wireless, 271

Cipher block chaining (CBC), 227–229, 231

Cochannel interference, 10

Code excited linear predictor (CELP), 12–13

Coding scheme-2, 17

Common Management Information Protocol (CMIP), 217

Consumer segment, 261

Counter-Mode with CBC-MAC Protocol (CCMP), 231–233

Recent Titles in the Artech House
Mobile Communications Series

John Walker, Series Editor

For further information on these and other Artech House titles, including previously considered out-of-print books now available through our In-Print-Forever® (IPF®) program, contact:

Artech House	Artech House
685 Canton Street	46 Gillingham Street
Norwood, MA 02062	London SW1V 1AH UK
Phone: 781-769-9750	Phone: +44 (0)20 7596-8750
Fax: 781-769-6334	Fax: +44 (0)20 7630-0166
e-mail: artech@artechhouse.com	e-mail: artech-uk@artechhouse.com

Find us on the World Wide Web at: www.artechhouse.com